Gerhard Zucker

Processing and Symbolization of Ambient Sensor Data

Gerhard Zucker

Processing and Symbolization of Ambient Sensor Data

A Neuropsychoanalytic Approach

Südwestdeutscher Verlag für Hochschulschriften

Impressum/Imprint (nur für Deutschland/ only for Germany)
Bibliografische Information der Deutschen Nationalbibliothek: Die Deutsche Nationalbibliothek verzeichnet diese Publikation in der Deutschen Nationalbibliografie; detaillierte bibliografische Daten sind im Internet über http://dnb.d-nb.de abrufbar.
Alle in diesem Buch genannten Marken und Produktnamen unterliegen warenzeichen-, marken- oder patentrechtlichem Schutz bzw. sind Warenzeichen oder eingetragene Warenzeichen der jeweiligen Inhaber. Die Wiedergabe von Marken, Produktnamen, Gebrauchsnamen, Handelsnamen, Warenbezeichnungen u.s.w. in diesem Werk berechtigt auch ohne besondere Kennzeichnung nicht zu der Annahme, dass solche Namen im Sinne der Warenzeichen- und Markenschutzgesetzgebung als frei zu betrachten wären und daher von jedermann benutzt werden dürften.

Verlag: Südwestdeutscher Verlag für Hochschulschriften Aktiengesellschaft & Co. KG
Dudweiler Landstr. 99, 66123 Saarbrücken, Deutschland
Telefon +49 681 37 20 271-1, Telefax +49 681 37 20 271-0, Email: info@svh-verlag.de
Zugl.: Vienna, Vienna University of Technology, Dissertation, 2006

Herstellung in Deutschland:
Schaltungsdienst Lange o.H.G., Berlin
Books on Demand GmbH, Norderstedt
Reha GmbH, Saarbrücken
Amazon Distribution GmbH, Leipzig
ISBN: 978-3-8381-1064-6

Imprint (only for USA, GB)
Bibliographic information published by the Deutsche Nationalbibliothek: The Deutsche Nationalbibliothek lists this publication in the Deutsche Nationalbibliografie; detailed bibliographic data are available in the Internet at http://dnb.d-nb.de.
Any brand names and product names mentioned in this book are subject to trademark, brand or patent protection and are trademarks or registered trademarks of their respective holders. The use of brand names, product names, common names, trade names, product descriptions etc. even without a particular marking in this works is in no way to be construed to mean that such names may be regarded as unrestricted in respect of trademark and brand protection legislation and could thus be used by anyone.

Publisher:
Südwestdeutscher Verlag für Hochschulschriften Aktiengesellschaft & Co. KG
Dudweiler Landstr. 99, 66123 Saarbrücken, Germany
Phone +49 681 37 20 271-1, Fax +49 681 37 20 271-0, Email: info@svh-verlag.de

Copyright © 2009 by the author and Südwestdeutscher Verlag für Hochschulschriften Aktiengesellschaft & Co. KG and licensors
All rights reserved. Saarbrücken 2009

Printed in the U.S.A.
Printed in the U.K. by (see last page)
ISBN: 978-3-8381-1064-6

Abstract

Building automation is a constantly growing domain that becomes more and more a part of our everyday life. Systems designed to support the human user will provide new types of services. A challenge for the design of such systems is the ability to observe objects, events and situations in a way that is similar to human perception.

This work defines a model for data processing based on the adaptation of neuroscientific, psychological and psychoanalytical models. By symbolizing data which originates from diverse types of sensors and processing it in multiple layers, relevant information is extracted and separated from unimportant data. This is achieved by applying knowledge that has been obtained from understanding how human perception operates. The result of this process is a symbolic representation of the world that is used to identify scenarios – sequences of events subject to time constraints. Because the focus of this work is on perception and surveillance, actions and reactions are of secondary significance, but are possible as a consequence of identified scenarios.

The symbols defined in this work are linked to the real world, in which the symbol operates – the symbols are grounded. This has been achieved by linking the lowest symbolic layer to real-world sensory data.

Preface

Technology has reached a level that allows for convenient integration of sensory equipment and provides sufficient computing resources at reasonable prices. Looking at today's setup of computational resources, an outside observer would expect machines with some basic human-like abilities that capable of increasing human safety and comfort. But while we have successfully taken pictures on the surface of Mars, the convenience of having a robot that prepares a cup of tea in an average kitchen is still not available. While machines excel humans by magnitudes when it comes to solving equations or processing big amounts of data, the solution to performing apparently "simple" tasks that are part of everyday human life still needs to be found.

This is the field of research to which this work attempts to contribute. Designing machines that support human users in a better way requires as a first step an understanding of the world as it is perceived by humans. Such machines need to operate in a natural environment designed for humans, but not for machines. Thus, we have to enable machines to recognize real-world objects, events and situations in a world of imprecise data, which so badly fit to the requirements of formal logic. Hence, it is necessary for such a system to perceive the real world, to classify and to identify. The goal of this process is a representation of the world that builds the base for higher-level functions.

Building machines that are able to fulfill tasks or even create something autonomously has been a personal goal for a long time and has motivated me to write this thesis. Sigmund Freud once stated that women suffer from penis envy – one of his lesser noteworthy theories. He claimed that at the age of about four girls start envying boys, because of their anatomical difference. A countermovement to this theory is something I can much better identify with: women are able to give birth to life and men, who lack this ability, have to compensate for it. One possible compensation is to concentrate on the design of technical systems, which imitate human behavior. My personal opinion is that this was one of my drives to become an engineer in the first place and to pick this topic for my thesis.

Looking at population development in European countries, there is reason to believe that taking care of the elderly will become more and more important in the future. People will live longer and thus require attention and caretaking during their retirement. Technical systems may increase comfort and safety for the elderly who live alone, by providing background observation and detecting possibly dangerous situations. One of the goals of this thesis is to aid these people with a system that provides safety, while respecting their privacy at the same time.

This thesis is embedded in the research activities of the Institute of Computer Technology and is based on the results of earlier works by colleagues and other researchers. The PhD theses by my former colleagues Gerhard Russ and Clara

Tamarit manifested my starting point for research activities. Over the years a considerable list of scientific publications developed, to which I contributed in one place or another. Three years ago, the project ARS was founded to unite the department's research; and it is this project, in which this work was created. Furthermore, my thesis is strongly influenced by the work of Mark Solms, Oliver Sacks, Antonio Damasio and Alexander Lurija, who all provided remarkable pieces of research in their fields and gave me an excellent starting point and further useful ideas for my work.

I want to thank my professor and supervisor Dietmar Dietrich who strongly contributed to this work with his previous research and with whom I had many good and fruitful discussions about this thesis. I also highly appreciate the work of my reviewers Brian O'Connell and Walter Penzhorn, who supported me especially in the last, most important phase of this work. My good friend Nicole Irmler rendered me a great service by applying the final touch to this work and I want to thank her for cross reading my thesis and removing many of the numerous typos and Germanisms.

I also extend my gratitude to my parents, who supported me during my time as a student. Finally and most important, I want to thank my wife Katharina Zucker, who filled me with love and motivation ever since we met – and especially while working on this thesis.

Table of Contents

Introduction 1
PART I: Prerequisites **5**
1   Biological Systems and the Human Mind 6
    1.1   The Mind as an emerging Result of Brain and Body 8
    1.2   Inner Representations 10
        1.2.1   Inner Representation of the Outer World 12
        1.2.2   Representation of the Inner World 13
    1.3   Memory and Knowledge 14
    1.4   Emotion 16
    1.5   Perception 17
        1.5.1   Extracting Constance and Relevance 19
        1.5.2   Identification and Classification 20
        1.5.3   Visual Perception 20
    1.6   Language 21
    1.7   Consciousness 22
2   Brain Mechanisms and Technical Counterparts 22
    2.1   Embodied Intelligence applied to a Building 27
    2.2   The Frame Problem in Artificial Intelligence 28
3   Symbolization 29
PART II: System Design **33**
4   Symbolic Processing 34
    4.1   Symbolic Model 35
        4.1.1   Symbol Hierarchy 37
        4.1.2   Symbolization in the Inner World 38
        4.1.3   Augmenting Symbols on Representation Level 39
    4.2   Symbolizing Physical Values 40
5   Sensory Equipment 41
    5.1   Light Barrier 42
    5.2   Motion Detector 43
    5.3   Tactile Sensor 44
    5.4   Temperature and Humidity Sensor 45
    5.5   Illumination Sensor 45
    5.6   Microphone 46

- 5.7 Camera 47
  - 5.7.1 Background Subtraction 47
  - 5.7.2 Face Detection 48
  - 5.7.3 Histograms 48
- 6 Knowledge and Memory 48
  - 6.1 Knowledge about the Environment 48
    - 6.1.1 Environment Layout 49
    - 6.1.2 Static Objects in the Environment 51
    - 6.1.3 Hierarchical Structure 51
  - 6.2 Associations 52
- 7 Perception and Its Prerequisites 54
  - 7.1 Location Information 54
  - 7.2 Object Recognition 55
    - 7.2.1 Object Feature Description 56
    - 7.2.2 Anchors 58
    - 7.2.3 Object Classification and Identification 58
- 8 World Representations 59
  - 8.1 Storage of Symbols and Historic Data 60
  - 8.2 Outer World Representation 62
  - 8.3 Inner World Representation 62
  - 8.4 Scenario Evaluation 63
  - 8.5 Action Subsystem 64
- 9 Reference Applications and Services 65
  - 9.1 Environment Model 66
  - 9.2 Application Service: Object Tracking 67
  - 9.3 Application Service: Person Tracking 67
  - 9.4 Application Service: Human Activities 68
  - 9.5 Application: Person Surveillance 70
  - 9.6 Application: Child Safety 71
  - 9.7 Application: Geriatric Care 72
  - 9.8 Application: Comfort 72
  - 9.9 Outline of Reference Applications 73
- **PART III: Concepts and Results** 77
- 10 Module Description and Data-Flow 77
  - 10.1 Modules for Outer World Representation 77

- 10.2 Modules for Inner World Representation 79
- 11 Symbol Definitions 80
  - 11.1 Symbol Categories and Naming Conventions 81
  - 11.2 Environment Model 82
  - 11.3 Application Service: Object Tracking 83
    - 11.3.1 Snapshot Symbols 84
    - 11.3.2 Representation Symbols 84
  - 11.4 Application Service: Person Tracking 85
    - 11.4.1 Sensors 85
    - 11.4.2 Microsymbols 86
    - 11.4.3 Snapshot Symbols 90
    - 11.4.4 Representation Symbols 92
  - 11.5 Application Service: Human Activities 92
    - 11.5.1 Activities 92
    - 11.5.2 Representation Symbols 93
  - 11.6 Application: Person Surveillance 93
  - 11.7 Application: Child Safety 93
    - 11.7.1 Scenario Symbols 94
    - 11.7.2 Action Series and Action Symbols 94
  - 11.8 Application: Geriatric Care 94
    - 11.8.1 Scenario Symbols 95
    - 11.8.2 Action Series and Action Symbols 95
  - 11.9 Application: Comfort 95
    - 11.9.1 Microsymbols 95
    - 11.9.2 Snapshot Symbols 95
    - 11.9.3 Representation Symbols 96
    - 11.9.4 Action Series Symbols 96
  - 11.10 Action Subsystem 96
  - 11.11 Inner World 96
    - 11.11.1 Microsymbols 97
    - 11.11.2 Representation Symbols 97
    - 11.11.3 Action Series Symbols 98
  - 11.12 Symbol Overview 98
- 12 Communications Design 100
  - 12.1 Integration between Field Level and WAN 100

- 12.2 Fieldbus Data Representation and Gathering 105
- 12.3 Communication Framework 107
  - 12.3.1 Message Format 108
  - 12.3.2 Communication Infrastructure 109
  - 12.3.3 Database Access 109
- 12.4 Database Storage 110
  - 12.4.1 Sensor Database Description 110
  - 12.4.2 Symbol Database Description 111
- 12.5 Time Representation 111
  - 12.5.1 Instant and Time Period 112
  - 12.5.2 Time as a Number 112
  - 12.5.3 Time Synchronization 116
- 13 Visualization, Simulation and Real-World Installation 116
  - 13.1 Visualization 117
  - 13.2 Simulation 118
  - 13.3 Smart Kitchen – A Real-World Installation 119
  - 13.4 Virtual Time and Real Time 119
  - 13.5 Simulation Run Generator 120
  - 13.6 Physical Model of Simulator 120
  - 13.7 Strategic Planning 120
- 14 Conclusion 121
- 15 Outlook 123
- Bibliography 125
- Internet Links 131

Introduction

Building automation has over the last decades matured to become an indispensable contribution to everyday life. This thesis shows a possibility to achieve new qualities in supporting humans by the use of technical systems that are capable of "understanding" the human world better.

While in former times only a few sensors and actuators were used by a system to interface with the real world, today we there are systems that use more and more sensors and actuators. Modern building automation systems already have some ten thousand sensors available. At the same time, the requirements have changed and new application fields were added: climate control should consider the presence of persons to improve energy management; the location of persons in a building is required for both safety and security; and it should be possible to protect persons from dangerous situations. Using more sensors provides a better view of the world, because more details are available. On the other hand, it becomes more important to distinguish between unimportant and important information and to not just get a process image, but a representation of the world as a human user would perceive it.

These goals require new models on higher data processing levels. There is a good source for such models, which has been used for decades and has recently experienced significant progress in research: the human mind. It is able to process a huge amount of information, extract the important facts and build a representation of the world. Recent research has brought forth new insights into human perception and the links between language and thought. Furthermore, consciousness has been separated into different levels of awareness. Finally, the bridge between neuroscience and psychoanalysis has been laid – two disciplines that were opposing each other for the longest time. Scientists now attempt to explain Sigmund Freud's theories using a sound neuropsychological basis.

The above mentioned rich source of new models is used in this thesis to provide the outlines of a system that is able to introduce new possibilities in building automation. Perceptive consciousness makes the world comprehensible for a machine, adapting it in a better way to the needs of human users.

Control engineering uses well-researched and precise processes that describe exactly how an optimal control loop has to be designed. It models a process, describes how the system reacts on variations of the inputs and uses this information to build a controller with defined transmission characteristics. As long as only a few inputs and outputs are necessary, this is indeed the best approach. However, in many systems this is not the case and it is necessary to use information from a lot of inputs to control a lot of outputs. Furthermore, the set of variables and configuration parameters cannot be properly modeled. When programming the control algorithm, a human programmer has to use knowledge about the process intuitively to get the best performing algorithm.

There is a lack of methods that allow handling complex input scenarios. These methods are needed to produce a stable system output independent of the current operating point and deal with nonlinear, redundant and maybe contradictory information. The parameters that are needed to control the process have to be comprehendible for a human user without having full understanding of the underlying process.

A common approach today is to analyze a problem and reduce it to significant parameters so that a machine without cognitive abilities is able to fulfill the task satisfactorily; instead of handling big amounts of partly redundant data, a model is designed that relies on a few variables, preferably only one. This is however not the approach of the system envisioned here.

The model that is described in this work builds the foundation for a system called ARS – *Artificial Recognition System*. In today's building automation systems, every sensor usually has its dedicated function in an application and is used exclusively for this application (e.g. a temperature sensor is used to measure the room temperature). Adding an additional sensor for the same functionality is not common, because it increases the costs of the system. In addition, it is also uncommon to use one sensor for different applications (e.g. a temperature sensors for heating control as well as for fire alarms). Such systems keep the overall complexity of the control circuitry low: if there is only one sensor to report the room temperature, it has to be taken as true; if there were more sensors, additional logic would be required to determine the most reasonable room temperature (in case the sensors report different temperatures). Disadvantages are, however, that the system has a limited view of the world that it operates in. An HVAC[1] system will perfectly adapt the room climate to meet human requirements, but it will do so no matter whether a person is actually present in the room or not; in case of fire, the HVAC system is unaware of it and might try to cool the room. Failure of a sensor results in malfunction that cannot be compensated for by other parts of the system. All types of system behavior have to be programmed into the system without it being able to adapt to modified requirements. Moreover, even if there were more sensors to provide additional information, it would be hard to benefit from this information due to the inability to handle the increased complexity.

The ARS system extends the existing concept of control systems, where a system is able to behave in a more sophisticated way. The goal is to have a system that perceives and reacts, based on diverse, redundant sensor information from different domains and industries – an ability that has until today not evolved too far [Lon02]. It shall extend the existing control loop approach by reacting upon personal needs of the user. The concepts introduced here are based on perceptive awareness (see [Tam03]

---

[1] Heating, Ventilation and Air Condition

and [Pra05b]) that is used to extract important information and abandon irrelevant information. Scenarios are recognized by using the concept of symbolization, for which foundations have been laid in [Rus03].

In the human mind, the lowest level of symbolization refers to the interface between the neuron level and the psyche. Symbolization is largely responsible for the way in which we see the real world. Neuroscientists have made detailed studies about the way the human brain works [Kap00] and the ARS system applies these results: instead of working with the information that each sensor provides, sensor information is condensed and processed as symbols in a multi-layer model (see also section 4.1). The system described here does not cover learning of new symbols (and associations between them), although the ability of learning is envisioned for a future system. The mechanisms and methods used here are predefined and do not change during the operation of the system. However, they are prepared to be adapted and modified, so that in a next step system-triggered modifications become possible.

Perceptive awareness is a combination of experience (which manifests as knowledge) and sensory updates reflecting the current state of the world. The information that is available from sensors is insufficient for a good world representation and has to be augmented by knowledge about the world. A human being learns these facts and rules during its lifetime and uses them intrinsically to build a world representation. Objects are classified, for example, as edible by their shape, size and color. This information is used to predict, for example, taste, structure and softness, relying on earlier experiences with similar objects that have taught us what to expect. Identification and classification of objects consider many different parameters and features (not just the few generic ones mentioned here). For most of them, we do not even have names. But it is this collection of knowledge and experience that enable us to build a world representation that contains far more information than just the data coming from our sensors. By applying predictions and hypotheses (based on earlier experience) to current sensory information, we are able to model the world and become aware of what is happening around us.

The ARS system is a supervisory system with limited possibilities to interact with the environment. It is intended as a support for human operators, meaning that it can constantly analyze incoming data and notify an operator about relevant scenarios that have been perceived. As a permanent observer of people and their activities, the ARS system can improve quality of life (while at the same time protecting privacy) and provide increased safety for people with special needs for supervision (e.g. children or the elderly).

PART I: Prerequisites

In Part I, the setting of this work is laid down. The different disciplines that are needed are presented by showing the results of past and present research. Related work is described and relevant concepts are examined. On the biological side, the results of analyzing and understanding different functionalities of the human brain is used for designing a model that allows the implementation of abilities of perceptive awareness into a system. Based on the understanding of the operation of the brain we have today, the attempt is made to apply this knowledge to a technical system. Algorithms and methods that are necessary to do so are identified in this part, laying the foundation for the second part of this work, where the system design is described.

Although the model for the system is taken from its biological counterpart, attempting to apply the knowledge gained in the fields of neurosciences, psychology and psychoanalysis, there are considerable differences between the motivations behind these two systems. Biological individuals follow certain needs that emerge from the requirements by the body to stay alive[2]. An individual will act in an egocentric way to gain benefit for itself (or at least for the group of individuals in which it is embedded); during most of its lifetime the main task is to survive, no matter what the consequences or implications for others are. A machine, on the other hand, is intended to support its human users and, for example, set human safety above its own needs. In the vision of the designer, such a machine will silently take care of itself (by fulfilling its basic needs) and otherwise modestly serve the human user without requesting anything but what is needed for it to operate correctly. Assuming that evolution "builds" humans and humans build machines, we see that there are differences in the requirements of "designing humans" versus designing machines.

This part of the thesis describes models of the human mind based on current research. There are two different possible approaches: either the brain is described with all its biological properties to get a detailed model of the brain as an organ. This first approach is useful when trying to understand human beings and the way they function. This work, however, tries to build a technical system with mind-like capabilities. Its goal is not to build models that describe the brain and its organic basis. Therefore, it is not sensible to describe the human mind on a biological basis; instead, the second approach is chosen, which is to extract models that can be used to build the envisioned system on a technical basis. In other words, the functionality of the human mind is migrated from a biological organ towards a technical system.

---

[2] Of course, this does not cover the full spectrum of motivations for human beings, but it still is a common denominator for humans and animals. On a very low but elementary level, the reason for existence of the brain and all its abilities (including consciousness) is to give the individual better chances to survive.

1 Biological Systems and the Human Mind

The transition between sensor level data and the symbolic processing methods is an important step that needs to be taken in order to complete this work. While neuropsychologists analyze – amongst other mechanisms – the operation of specialized neurons in the brain to get an understanding of their functionality, technical system cannot rely on the abilities of neurons. Instead, they implement existing algorithms and methods that originate from, for example, image processing. The ability of neurons in the optical cortex to detect edges (section 1.5.3) is substituted by edge detection algorithms known from image processing. A candidate for extracting the invariant features of different objects that belong to the same category (which is an ability of human visual perception) are artificial neural networks [And95]. Classification, that is, assigning perceptions to different categories, can be done in Fuzzy Logic [Zad65]. Information technology has brought forth a lot of different tools that can today be used to implement the behavior of human perception. Neurobiology has also made great progress in the last few years. New technologies allow studying brain activities in a much more detailed way than before. Human perception does not rely purely on sensor data, but depends to a great deal on memories, knowledge and the representation of the incoming data (see also [Sac87]). A lot of the mechanisms in the brain have been deciphered by analysis of brain scans, by studying electrochemical reactions in the brain and running test sequences with persons suffering from brain damage. Neuropsychologists like Mark Solms study the functionality of the brain on patients with lesions – damages of the brain, where the location is exactly known and usually limited to a certain area of the brain. The human mind is analyzed by observing the changes when different areas of the brain are damaged.

Furthermore, there is psychoanalysis, a discipline that is founded on the research of Sigmund Freud (1856-1939), who helped mentally ill people by using techniques previously unheard of: letting the patient talk, using free associations or interpreting dreams. He developed the model of three psychic instances: It, Ego and Superego.

Although Freud's method of psychoanalysis had been successful and up to the 1950s, it started being criticized when neurobiology could show successes in understanding the functionality of the brain [Lak05]. Using new methods, the human brain appeared to be understandable like any other organ of the human body. Since the brain is the carrier of the psyche, it appeared reasonable to influence the human mind by influencing its main organ, the brain. Big parts of perception were understood by scientists; memories and emotions could be tracked down to molecular level. Freud's theoretical constructs on the other hand were complex and impossible to put on a sound scientific basis that is grounded in experimental proof[3]. The big success that Freud and his followers had with psychoanalysis turned into the opposite [Lak05].

Only recently, neuroscience and psychoanalysis have made combined efforts[4] to explain the phenomena that make human beings special amongst other animals. Some of Freud's concepts are finally supported by neuroscientific observations: Mark Solms gives examples of how the two disciplines explain the same things in different languages [Lak05]. For instance, Mrs. Jacobs, one of his patients had a stroke that rendered the left part of her body useless. She could move neither arm nor leg and could not get up, not an uncommon handicap after having had a stroke. What is however uncommon is the fact that Mrs. Jacobs claims to be perfectly alright, she wants to go home, do her work and not be bothered by the annoying doctor. When being asked to move her left arm, it does not move, of course, but Mrs. Jacobs still claims that it did – at least in her head[5]. The syndrome is called *anosognosia* – not being aware of a physical handicap or disease – and was first described at the beginning of the 20th century by Babinsky [Bab14]. It seems as if the conscious mind of a person protects itself from a reality that might be too much to bear. Consciousness refuses to accept the obvious, because the truth is too much to take, instead, another truth is created. Explanations for this new truth are found so that the integrity of the individual can remain at least partly intact without having to cope with the blow of realities' impact.

While this syndrome cannot be fully explained by neuroscientific methods, the answer can be found in Freud's model: Mrs. Jacobs *represses* the fact that she is handicapped. What has long been known in psychoanalysis as repression can now be mapped onto neuropsychologic phenomena. Other cross-links have also been identified: Freud stated that the first few years are fundamental for the development of individuals, which is explained by neuroscience in that the limbic system is structurally modified during that time. Neuroscience explains why it is impossible to remember anything from the time of our earliest childhood, at least consciously, since this would require structures that are not fully completed at early childhood; Freud also understood this phenomenon. And the drives – one of the most fundamental concepts of Freud – can be identified today as systems in the brain, parallels between the messenger substance Dopamin and what Freud called *libido* are found [Lak05].

---

[3] At least not in the classical ways of natural science, where experiments need to be deterministic and reproducible. In other words, it is highly unlikely for psychoanalysis to meet the requirements of classical laboratory conditions, where there is a limited set of variables within a defined environment.

[4] An example for this cooperation is the journal "Neuro-psychoanalysis", which reports regularly about latest developments and contains commentaries from people working in the field. Its goal is to "*create an ongoing dialogue with the aim of reconciling psychoanalytic and neuroscientific perspectives on the mind*" [Neu06] thus bringing together the two disciplines that have been departed for the last decades.

[5] Other patients find plausible explanations why a specific body part would not actually move (e.g., they claim to have a stiff shoulder or arthritis).

1.1 The Mind as an emerging Result of Brain and Body

Different approaches have been made to explain the human ability to "see" things, meaning consciously perceiving objects and events. An early idea was that the brain receives images of the outside world and operates on these images. However, when the mind uses images, who looks at these images to understand their meaning? This resulted in the *homunculus problem*: the ability of persons to see things is explained by an inner mechanism that is responsible for the act of seeing[6]. The term homunculus is borrowed from the alchemist Paracelsus[7]. At the end of the 15th century, the alchemist and doctor Paracelsus (1493-1541) claimed to have created a "homunculus" – an artificial human being made of flesh and blood [Nus04]. Therefore, the conscious perception, which had to be explained in some manner, is actually done by a little man that sits in everybody's head. By explaining the mechanism in this way, the problem is deferred to another entity, where the question arises, how the homunculus itself works.

In search for an adequate model to describe the ability of beings to interact with its environment, R.A. Brooks has introduced the *subsumption architecture* [Bro86] to describe the behavior of low animals such as insects. In this architecture, a perception directly causes an action without any cognitive modification in between. This means that the action directly depends on the stimulus and cannot be modified through the lifetime of the individual. The architecture consists of modules that are modeled as finite state machines, a property, which is very convenient for technical implementation. These modules are organized in layers, which implement certain behavior (e.g. walk around, pursue object) and therefore build a hierarchical architecture, since a higher layer can inhibit lower layers. Layers are functional on their own, meaning that the layer responsible for basic movements and reflexes is still operative if higher layers are disabled. Because the finite state machines are always running, the architecture is asynchronous and parallel: asynchronous, because there is no need to synchronize different modules with each other – higher layers simply inhibit lower layer tasks at any given time and in parallel. Each finite state machine runs its own, separate execution without being aware of other modules (with the exception of inhibition).

A big disadvantage of subsumption architecture for the purposes of this work is the fact that it has no symbolic knowledge representation and no world model. On the other hand, this makes the architecture very fast – i.e. a sensory input causes a reaction immediately – and it is possible to build simple systems that are able to deal with a complex environment. But a system built using subsumption architecture is purely reactive and cannot plan actions for the near or far future. The lack of a world

---

[6] An early, enchanted attempt to explain what can be compared to Damasios *hard problem* of consciousness [Dam99].

[7] His actual name was Philippus Aureolus Theophrastus Bombast von Hohenheim.

model makes it impossible for the system to understand its environment and thus does not allow creating strategies. Since there is no symbolic representation of knowledge or the environment, there is no way for the system to abstract from sensory inputs or to reason about its behavior.

Seen from an evolutionary standpoint it appears reasonable that an architecture similar to subsumption architecture is still the base for human interaction with the outer world. Reflexes are a part of human behavior and a hierarchical model is fairly reasonable for the lower levels of human actions and reactions. However, higher cognitive functions are not possible in a system based solely on subsumption architecture. The human mind remains the system of interest for this work.

The connections and dependencies between different brain areas have been subject to permanent research in the last decades. The link between the brain as an organ and the human mind can today be understood from different sides.

Recent research proposes, for example, that depressions can not only be treated by psychologists (i.e. by operating on the human mind), but find their solution in a well-targeted medical treatment: [Pez05] shows that the brain-circuitry between cingulate and amygdale, two brain structures responsible for balancing feelings, can be influenced by psychiatric drugs – with lasting effect. Long sessions with psychologists are not the only way to modify connections between brain cells permanently, but also a medical treatment. Depressions are treated from two different sides, by treating the organ and by treating the mind. Again, we see two disciplines merging: some diseases can be treated better by medication, some by psychologists.

In early attempts to understand the brain, a very persuading viewpoint was localizationism, which assigns brain regions to functions of the brain. The goal is to identify a cluster of neurons and refer to it as the region, which is responsible for certain functionality. Although this is possible for regions that were inherited from our evolutionary ancestors and are responsible for low-level functions, the attempts made by various scientists went far beyond simple functions and were conducted mainly by intuition – and usually lacked any sound prove. The assumption that every function of the brain can be isolated in a brain region has no relevance any more: it is true for basic function (e.g. edge detection in the optical cortex [Gol02]), but not for any kind of higher function, especially not for cognitive functions. Intentional movements and manipulation require a complex functional system. Digestion and respiration are also functional systems, which possess a goal; in case of respiration, it is supply of oxygen for alveoli in the lounge. The way this goal is achieved depends on the control of different body muscles and is therefore a complex system. The control of this system cannot be located in one specific brain region. Lurija proposes to move away from localization of functions and to not localize higher mental processes, but find out, which groups of brain areas cooperate [Lur01]. This "systemic localization" manifests in the systems of cooperating brain areas, which may be physically located next to each other, but can also be cooperating over a large distance. In case of higher mental functions, the development over the lifetime of the individual is also an important factor. Higher mental functions are not static: first a mental function (e.g. writing) is expansive and uses external helpers; only after a lot of practice, the function turns into automated motor actions [Lur01].

1.2 Inner Representations

Humankind has for a long time pondered the question how the world really is, what really lies outside of our own body. The tools we have at our disposal at first hand are the sensors we are equipped with e.g. seeing, hearing, smelling or somatic sensors. These sensors are the interface to the outer world and are the first level of what we can perceive as being the real, physical world outside. When we start exploring the world while we are still toddlers, we gain a lot of knowledge about the things that surround us. Although we are not aware of the laws of physics, we still learn how the world works by interacting with it: things fall to the ground, some of them break, and some do not. The process of learning continues through our life and we refine our knowledge, which we can increasingly use to deduce facts that we have not experienced ourselves in the first place. By having experienced that a glass vase breaks when it falls on the floor we can predict that a drinking glass will do the same without actually having to break it.

After years of experiencing the same physical world it becomes more and more predictable to us, hardly anything surprises us, and if it does, we usually find a good explanation for it later on. Sometimes glass is made in such a way that it is resistant to a certain level of physical blows and therefore might not break. This deep knowledge of a big amount of different facts and relations encourages us to believe that we know what the physical world really is. The senses we are equipped with do not betray us. Moreover, since humankind has passed a long line of evolutionary steps in its long development this is in fact true: the sensors we have are very well adapted to the needs we have and most of the time match what can be shown to be the real, physical world. Sometimes this does not work, as can be shown by various optical illusions, where on the one hand our eyes tell us one set of facts, while we can use other tools (and our mind) to prove that it can actually not be true. These illusions can tell us a lot about how human perception and it ways of operation.

Still, there are physical effects and forces in the physical world that we are unable to perceive. Nobody can tell whether an electric wire has a (low) voltage without using additional tools (higher voltages would cause pain and would therefore be perceivable). The same applies to radioactive radiation. We are not equipped with proper sensors to detect electric voltage or radiation – it was not necessary during our evolution. The world we perceive is only a subset of the real, physical world, the outer world. But this limited view of the world is not only true for physical effects that we are unable to perceive by sensors. During the personal development that every human being goes through, we collect pieces of facts and relations that we gain from the outer world. Children are very good at ignoring facts that they are unable to comprehend. They are robust against overloading them with too much information (it is still possible, though). Whatever cannot be understood is ignored. Children have limited knowledge about the world they live in and are still able to handle whatever is important for them to survive and evolve.

This does not change when we grow up, only the amount of facts and relations we know about the outer world becomes more fine-grained. Still it is important to understand that there is an outer world, which we explore to create our own view inside of us. This *representation of the outer world* matches to a big extent the effects we perceive and allows us to reliably interact with the outer world. But the representation of the outer world and the outer world itself are not identical. Even more, the inner representation is different for every human being, since everybody has his own history of gathering facts and experiences in the outer world. And, the inner representation of the outer world is only a subset of the outer world.

To sum up, the inner representation of the outer world is an image of the real world, which to a great extent matches the real world, but has its own ways of representing the world. Comprehension results from perceiving events and deducing relations, without necessarily fully understanding the processes. A child can understand magnetism as an effect where certain (metallic) objects are attracted to each other, but there is no need to understand the laws of magnetism to do so. Therefore, the outer world mechanism "magnetism" is translated to the inner representation merely by its effect.

A human being does not only have the outer world in which it is embedded, it also has an "inner world", which is the internal milieu of the body [Sol02]. In the early days of humankind, the brain was the organ to fulfill the needs of the inner world by interacting with the outer world. To keep the body alive, the individual had to get supplies from the outer world in terms of oxygen, water and food but also had to obey a broad variety of other internal parameters like body temperature, heart rate, digestion and so on. The brain was always in the middle between the two worlds and had to translate between them. It is important is to see that both worlds have their own representations in the brain. The representation of the outer world gives us a "world model" of the physical environment around us, while the *representation of the inner world* contains the current status of our body, its needs, pains or requirements.

The question arises how these representations can be mapped to a machine that shall perceive its environment and what the abilities of such a system can be. Similar to humans, the system has a representation of the outer world. Certainly, this representation is limited and not able to compete with the one of grown up humans. It consists only of facts and relations that are important for the system to fulfill its tasks[8]. In this matter, the abilities of the system are closer to that of a handicapped

---

[8] The tasks that have been chosen as reference applications are described in section 9.

person, such as Kaspar Hauser [Hau95]. Hauser was found in 1828 on the streets Nuremberg in a disoriented state of mind. From what could be constructed, he had to spend his childhood in a cellar, isolated from the rest of the world. His vocabulary was limited in the beginning and contained only about 50 words. However, he was very good at comprehending and learned to read and write before he died in 1833. Kaspar Hauser was in the beginning only able to understand a limited set of facts, since he had no possibility to explore the world as we know it. Everything was new to him and therefore overstrained his abilities. He had to watch, explore and understand this new world in the same way as children do it. Only after he was acquainted with parts of this new world he was able to behave more and more in a way that we would call "normal". The ARS system only comprehends a limited subset of the understanding of a human. Still it is not child-like, since it possesses high-level cognitive functions. The inner representation of the outer world in the system is limited to the necessary facts and relations.

1.2.1 Inner Representation of the Outer World

As said before, there is an outer world, the real, physical world; by perceiving it we build a representation of this world, the *inner representation of the outer world*[9]. A representation of the world is necessary, because a system without such a representation cannot implement cognitive functionality (see section 1.1). The inner representation of the outer world contains the world model – what the world is like when looked at from one specific individual. The ARS system uses this principle and creates a representation of the outer world in order to interact with the real world and understand the events that take place.

In the human brain, the sensory information is projected onto *body maps*, which are areas in the cortex. Since the location of the sensor influences the position where the information is projected to, the spatial information of the original sensor location is maintained and thus available in the cortex. For example, the lower-left corner of the eyes' retina is always mapped onto the lower half of the visual projection cortex. Similarly, the whole surface of the body can be found as body maps in the brain, where somatic sensations are projected onto the primary sensory cortex of the parietal lobe [Sol02][10]. On a higher functional level above these body maps the *association cortex* contains functions like object recognition. It is responsible for collecting information and integrating the information into something that is actually recognized. Recognizing a car, for example, can be done by various different views from different sides, but it can also be perceived by mere sound without any visual information. Each of the visual impressions and all of the different sounds a car can

---

[9] For reasons of brevity this representation is also called *representation of the outer world*, see also section 1.2.2.

[10] Wilder Penfield created a drawing of these projections, which he humorously called homunculus – the little man inside a person [Pen50].

make contribute to realizing the fact that there is a car in the outside world. The association cortex (with its many sub areas) is the area in the brain where the information is integrated and combined into the representation of one object: the car. If we wanted to identify an area in the brain where we would look for a representation of the outer world, this would be it – but only in a very general sense. There have been attempts to pin down sensations like the above-mentioned car onto narrow regions or even single neurons in the brain, but they were doomed to fail. Just like the calculation in the erroneous Pentium processor (see chapter 2) cannot be described by pointing out a set of transistors that are responsible for doing the calculation and then identifying the faulty transistor, it is not possible to find the world representation or even single thoughts about the world in the cortex. The calculation in the processor is a sequence of commands, where each commands uses different areas of the processor to execute. After the whole sequence has been executed, the result is available, but it was caused by a temporal sequence applied to different units of the processor. The brain, on the other side, needs all its different areas to create the world representation and – on even higher layers – allows us to reflect consciously on this world, to abstract and to think in an abstracted manner. Calculations in a processor are not a state, they are a process – and so is the human mind.

The inner representation of the outer world is a combination of the sensory information (after they have been processed by other areas) and knowledge about the world. Only after learning the different sounds and images of a car, we are able to identify just a single sound and deduce the existence of a car (which may currently not be visible to us). The representation is therefore based on sensory information, which is augmented by previously learnt knowledge. In a way, the representation that we think of being the real world is merely a construction of past memories that are used to give meaning to the current sensor information. We do not see the real world, but rather our own creation – and because there can be more than one interpretation, there can be more than one world representation and we chose the one that appears most reasonable to us[11].

1.2.2 Representation of the Inner World

The brain as the organ between the inner and outer world has information from its sensors that it uses to construct the representation of the outer world. Based on this representation the individual can interact with the real world. Similarly, the brain also has to have information from the body itself and to know about respiration, blood pressure, body temperature and so on so that it can act according to the needs of the body. In the language of Freud, these needs are the *drives* that emerge from our bodily needs. Modifications of drives are then experienced as *emotions* [Sol02]. Again, there first needs to be a representation of the inner milieu, the *representation of the inner world*[12]. Information about the inner milieu is registered in the hypothalamus; the functional unit responsible for creating a representation of the inner world – if we would want to look for it in the brain – can be identified as the limbic system together with the frontal cortex [Sol02]. As opposed to the representation of the outer world, there is no ambiguity in the representation of the inner world, simply because there is only one mind in one body, thus there cannot be different opinions like in the real world. Still, it does not necessarily mean that the mind knows the one and only truth about the body. The origin of a pain can often be misinterpreted and might need medical assistance from an expert to be located correctly. In any case, the body will react to the pain in the best way it can – may it be good or not.

The inner milieu and thus its representation in the mind dictates the course of action to be taken. This is very important for a biological system, since the correct operation of the body is vital for the existence of the individual, but it does not play such an important role for a technical system (which can be rebooted or repaired). Still the system can be separated into the outer world that is not part of the system and the components that belong to it. This way it is possible to define a "body" of the system (however peculiar it might look) and include the state of this body into the representation of the inner world.

1.3 Memory and Knowledge

Memory and its reconstruction is a complex process in the human mind. Aside of the broad amount of information that is not consciously accessible, but still influences our everyday decisions and reactions to events (see section 1.4), there is the autobiographic self, which combines events with the individuals self experience. This *episodic memory* allows a person to recall incidents that have happened to herself or himself [Sol02]. Recalling here means to consciously reconstruct the memory of this event.

Other classes of memory are *semantic memory* and *procedural memory* [Sol02]. The most interesting one for the scope of this work is semantic memory. It contains the knowledge about the world and its mechanisms. It is a collection of bits of information that make up our understanding of the world in objective information. It contains facts and categories, meanings of words as well as grammatical rules, mathematic values and knowledge of shapes. Unlike episodic memory the

---

[11] The examples in section 1.5 show that our construction of the world can differ from the construction of other people and that what we understand depends on what we know.

[12] Note that the term "representation of the inner world" instead of the longer term "inner representation of the inner world" is used (see also section 1.2.1). While sometimes the term "inner representation of the outside world", will be used in this work, there is no need to stress this for the inner world, since it is an internal representation of internal impressions and will not be represented anywhere else.

information is not related to the individual, but is objective, seen from a third-person view [Sol02] (e.g. "Fire is hot", "Vienna is the capitol of Austria", and so on).

Procedural memory contains motor skills that enable a person to fulfill tasks like walking or grabbing an object. Everything that requires movement, but is done without being controlled consciously is part of procedural memory. By continuous repetition of a new skill, it becomes a habitual behavior that can be executed unconsciously [Sol02].

Memories are laid down in the mind in various different ways and are thus accessible differently. A strong difference between memory in the technical sense and human memory is that human memory cannot be located and, for example, destroyed easily: it is a known fact that the longer a memory is stored, the better it is protected against loss. On the basis of clinical observations, Ribot observed that brain injury affects memories in the reverse order of their formation. Newly gained memories are most vulnerable against brain damages, while old memory is hardly destructible. It seems that over time memory is more and more engraved into the brain, making deeper cuts the longer it is around. This has become known as *Ribot's Law* [Scl02].

Scientists today also differ between *short-term memory* and *long-term memory*[13]. Short-term memory is also called *working memory* [Sol02], since it refers to facts and events that the brain can operate on. After a few seconds, memories are consolidated and become part of long-term memory. While it appears that most of what we experience during the day is lost shortly after and not available any more, it may well be that this is not the case. Sigmund Freud stated:

> *Perhaps we ought to content ourselves with asserting that what is past mental life may be preserved and is not necessarily destroyed. It is always possible that even in the mind some of what is old is effaced or abscrbed... to such an extent that it cannot be restored or revived by any means; or that preservation in general is dependent on certain favourable conditions. It is possible, but we know nothing about it.* [Fre74]

The question is, whether an individual is able to consciously reconstruct events that have happened or if conscious access is not available[14]. The memory itself is still present and influences the individual's behavior and perception of the world. Therefore, memory is very resistant to forgetting, especially the longer it has been around. But not necessarily all aspects are available. Metaphorically speaking: when you dig a hole in the ground, the hole will still be there after the act of digging; the ground will remember the hole. Who dug the hole and with which tool, however, may

---

[13] Where short-term spans over the timeframe of a few seconds; anything that is longer ago is already referred to as long-term memory.

[14] At least for the types of memory that can be accessed consciously, namely episodic memory.

well be forgotten. Models for the human way of storing memories have not advanced very far, at least in terms of how to apply them to technical implementations. The human memory is a complex construction that is tightly woven into all the capabilities of the brain. Unlike the current memory architecture in computers, where memory and data processing can be separated from each other, the architecture of human memory merges storage and processing into one unit.

Perception appears to be an objective process, where information from the outside world is used to update the view an individual has of the world. However, it has been shown that perceptions strongly relies on memory [Sol02]. What is not known cannot be perceived[15]. Solms [Sol02] gives an example of a cat that has been deprived from the ability to see horizontal lines. Consequently, it does not perceive horizontal lines (and bumps into any horizontal bar that is in its way). In fact, what we perceive as the world surrounding us consists to a great extent of memories of a past world as we have experienced it earlier. Identification and categorization (section 1.5.2) rely on categories that are learned and stored in memory. Once an apple is identified as an apple, perception can be reduced to merely acknowledging that nothing has been found that contradicts against the believe that the object at hand is actually an apple. Identification is similar, which can be experienced in everyday life when, for example, not registering a new haircut of a familiar person. The image of the person is not refreshed; instead, the old image from memory is used.

1.4 Emotion

The human brain is equipped with various subcortical emotion systems and the current state of research suggests that there is an interconnection between these emotions and the cognitive systems of the frontal lobes [Sol02]. Experiments have shown that the judgmental abilities of humans strongly depend on emotions. In the Iowa Gambling Task [Bec94] persons take cards of different decks and win or lose money on every card depending on complex game rules. Although the persons do not comprehend the game in its full complexity, a healthy person (meaning a person without brain damages) is soon able to tell, which of the decks is good and which is bad. What is described as "a hunch" or "a good feeling" is actually the information coming from a second, non-cognitive source of information, which attaches emotions to the card decks. Decisions are not based on analytical, cognitive evaluation of the game, but rather on affective level [Sol02].

Human beings have two possibilities to evaluate situations: either by using cognitive analysis of facts that can be formulated and consciously operated on or by emotion [Dam94]. While it appears infeasible to let a decision or hypothesis be guided by an emotion rather than by thinking it through, the two sources of information are not so

---

[15] It can, however, be learned and thus become known. The above statement is therefore not true for humans, but still applies to the ARS system envisioned in this work.

far separated from each other. [Ryd04] proposes a theory of mental representation by introducing a neuronal model. One of the main abilities of neurons is to extract hidden environmental variables by combining sensory input. This input contains the hidden variable in different ways; Ryder claims that neurons are able to extract this redundant information (which is not directly visible as sensory input). By connecting more layers of neurons, where each layer is connected to the previous one, it should be possible to extract more complex hidden variables from sensory information.

If this concept is extended, the result is a judgment of a situation that may be beyond conscious understanding, because the process of extracting hidden variables is not consciously accessible. Still, the layers of neurons attach an emotion to a situation that can be used for making a decision (as happened in the Iowa Gambling Task). Emotions can then be gained by extracting hidden environmental variables. If these emotions are stored in memory and can later be retrieved when perceiving a situation or an object, a fast judgment of a situation or object appears possible. Unfortunately, little is known so far about the actual "implementation" of such brain structures, therefore it is not considered relevant for direct application to a technical system and is thus not in the scope of this thesis.

1.5 Perception

Processing input that originates from the outside physical world is a key issue for a system that shall operate in an environment designed for humans and not for machines. A lot of information is available to the system by evaluating the data that comes from the various different sensors. The task of processing information from the outer world and creating some kind of inner representation is done by *perception*. Today every such system has some means to process sensor data and react accordingly. However, when the number of data increases, the strictly algorithmic methods become too complex to be manageable [Pra05a]. Perception as it is meant here works differently. Scientists have been studying perception in human beings and animals for a long time. In the 19th century, visual perception was understood as a passive stimulation of the retina and the optical cortex. The stimulus from the outside is reproduced in the visual cortex and the generated structure is thus isomorphic to the stimulus. The structure of an object that a person looks at creates a stimulus pattern that reflects this structure. This was the idea of *isomorphism* [Lur01].

Today psychology interprets perception as a process, which extracts features from the available information; perception is an active process that seeks information, builds hypotheses, compares extracted characteristic features and reaches a point where the hypotheses match the available data [Lur01]. The brain permanently faces a vast amount of information from the body's sensory equipment. This information has to be separated, synthesized and passed on to the responsible subsystems, which are specialized for certain subtasks in perception. A significant property of human perception is that objects, facts and events that are well known are perceived more easily than those, which are completely unknown. Perception – like many other abilities of humans – is improved over time by shortening the patterns that are necessary to fulfill a task[16].

A key issue in perception is the coding of information that can be processed by the brain. The simple mechanistic model of the 19th century is unable to explain the mechanisms that are necessary to process the amount of information that is necessary for human perception. A much better approach is to use a hierarchical model that groups pieces of information together and translates sensory data into codes that build the base for higher-level systems.

A good idea of how human perception works is by looking at situations where the well-tuned apparatus of perceptions fails. This is, for example, the case of visual illusions (see [Max02]) or when misunderstanding spoken words and especially when listening to lyrics of a song. The act of mishearing text is well known throughout different countries, it even has become a technical term in English speaking parts of the world: a *Mondegreen* [Hac04]. This term has been coined by the American writer Sylvia Wright. When she was a child, she listened to the Scottish ballad "The Bonny Earl of Murray" where she understood:

> *They ha'e slain the Earl of Murray / And Lady Mondegreen.*

While the original lyrics of the ballad are as follows:

> *They ha'e slain the Earl of Murray / And laid him on the green.*

The name of Lady Mondegreen – which never occurred in the original text – thus describes misunderstanding of text spoken or sung. Such mistakes are common and show us that understanding and perception rely not only on the available data, but also on knowledge. Another example shall show how much this knowledge influences perception. In 1778, Matthias Claudius created "Der Mond ist aufgegangen", a poem that was later set to music in a song called "Abendlied". The first verse contains the lines

> *Der Wald steht schwarz und schweiget*
>
> *Und aus den Wiesen steiget*
>
> *Der weiße Nebel wunderbar*[17]

---

[16] The same is true for, for example, reading or writing. Not only perceptional, but also motor tasks are accelerated, when done frequently. In the beginning, handwriting is compiled of a sequence of single motor impulses. After continuous practice, handwriting turns into a uniform movement that does not require drawing each character separately [Lur01].

[17] It translates to "The forest stands dark and silent / And from the meadows climbs / A wondrous white mist".

The second line was misunderstood by children living in Munich, who turned it into "Und aus der Isar steiget" [Hac04], which means "and from the Isar climbs".

This example shows how perception is a construction process rather than a passive mechanism for processing incoming data. The song line is matched to something that is as close as possible to known words or facts. Since the river Isar runs through Munich, it is understandable that children living in Munich find the most reasonable match for the text line the well-known river rather than some meadows.

The same applies to other situations where perception is applied. Only that most of the time experience ensures that most of the people observing a situation have similar enough knowledge that the selected hypothesis of the situation (or the inner representation of the outer world, to be more precise) overlaps to a great extent between different persons. This is what makes interaction between human beings in a society possible: a common agreement on a big amount of facts. On the other hand, it makes the task of actual classification difficult: since the physical world in which we are embedded does not contain clear boundaries between different classes of fact, events or objects, each classification[18] will contain (or leave out) something on which different persons will not reach a common agreement. A chair is a commonly understood object, but classifying a given object as being a chair or a stool will most likely not yield an unambiguous result when asking different persons.

1.5.1 Extracting Constance and Relevance

The real-world environment in which human beings are embedded (and which is the environment of the system envisioned here), contains rich sources of information for different types of sensors: visual and aural perceptions are the richest sources, but also physical parameters like, for example, humidity and temperature provide important contributions. While it is today possible to build sensors that can pick up all kinds of information, the question arises how to further process this vast amount of information.

Lurija describes perception as a process that looks for information, identifies characteristic features, compares it to stored memories and builds a hypothesis [Lur01]. The comparison with formerly learned and thus known perceptions requires the information to be decomposed using codes[19]. Visual perception has been well researched and so we know today that visual analysis in the optical cortex is done by a big amount of neurons, where each neuron reacts on a specific feature of the perceived object. The process of perception relies on memory (as described in section 1.3), everything that is known can be perceived quickly, only when something is new, we have use our full consciousness to study and experience it. Perception is never

---

[18] The term classification is meant as the definition of boundaries between different classes of facts, events or objects; other authors refer to it as discrimination [Har90].

[19] These codes build the base for what is later referred to as *symbols* in this work.

complete, it gives an impression of the surrounding world, but is not able to grasp it completely (this would, for example, require looking at an object from all side, before drawing the conclusion what the object is). Therefore, perception is used to give use clues about what is happening in the outside world; these clues are then combined with knowledge about the world (semantic memory, see section 1.3) to create a world representation. Due to the abilities of the human brain this world representation is usually a very close match to the objective, physical world and it is a great challenge for any technical system to achieve anything similar. Well known optical illusions reveal the "tricks" that the human brain uses to create this representation, but when perceiving the environment that commonly surrounds us, hardly any mistakes are made.

1.5.2 Identification and Classification

According to Harnad [Har90], human beings can "discriminate, manipulate, identify and describe the objects, events and states of affairs in the world they live in". The two most important abilities for the ARS system are "discriminate" and "identify". Discrimination, or classification, is the act of assigning a level of similarity to sensory input, thus being able to tell that two inputs belong to the same class. Identification refers to giving a "name" to a class of inputs, that is, being able to treat inputs as belonging to the same class. Both acts require the ability to extract constant features from the vast amount of information that is available. Assigning an identifier is the beginning of creating a language that can be used for higher-level cognitive processes (section 1.6).

Using what has been said in section 1.5.1, classification builds hypotheses to find proper classes for an input. If further inputs strengthen one of the hypotheses, it becomes the favored one. Similarly, well-known objects or events are backed by more hypotheses that allow to classify the object faster with only limited input.

1.5.3 Visual Perception

Visual perception in the human brain has on its lowest level, i.e. closest to the visual sensors, the task of feature extraction and pattern recognition. On the way from the eye to the visual cortex, specialized brain cells extract various types of information. Visual impressions can be seen as a set of lines and shapes that need to be processed. Aside of color information, cortical cells are able to detect orientation. [Gol02] shows a cortex cell that has a receptive field, which responds to vertically aligned bars by changing the impulses per second that are fired: only if orientation is perfectly vertical, the cell fires with about 25 impulses per second, changes in the angle result in lower firing rates. Additionally, hypercomplex cells can detect corners, angles and discontinuity [Gol02]. Depending on the alignment of cells with similar orientation preference, the brain can combine a bigger picture from the activity patterns of the neurons. Recent research [Wol05] has explained this alignment by a self-organizing process that produces a different architecture for every individual. Locations that are close to each other in the visual field are coded by neurons that are also close to each other in the visual cortex[20] [Gol02] (see also the description of body maps in section 1.2.1).

The specialization of neurons for tasks like detecting movement, edges or shapes has a long evolutionary history and is as such present in animals as well. For example, in [Wan92] it is shown that pigeons have the ability to detect rapidly approaching objects (which indicate possible danger) by the use of neurons that respond selectively to objects moving on a collision course towards the bird.

1.6 Language

Nothing is more convenient than having a word for a thing or an event. A description in spoken language is very precise compared to the internal states of the mind that also describe this thing or event. However, the "language" referred to in this section goes beyond the concept of spoken or written language. Spoken language is a subset of the means of communication that operate in the human mind. A lot of facts or judgments are not easily put into spoken words, but still contribute to perception as well as to judgment (section 1.4).

Language is today seen as the key to consciousness [Sol02] and it is understood that language and consciousness if not dependent on each other at least share a common base [Lur01]. Many of the different properties in understanding and speaking language are analyzed by lesions, that is, localized damages in the brain [Sol02]. Lesions in different brain regions cause a patient to be unable to pronounce a word, although they know the meaning of it, or the opposite: a word is heard, but the meaning cannot be retrieved. Other damages make it impossible for patients to grasp whole sentences, although the words are understood.

Because the "internal description" is much more complex and therefore contains a lot more details and cross connections than the spoken language, operating on spoken terms leads to more precise results and often is the only way to achieve a result at all. An example shall underpin this: a child learned about basic geometry in school, two-dimensional shapes and in particular the right-angled triangle. Before learning more about the correlations between the lengths of sides in the right-angled triangle, the child was already pondering the question of relationships between the sides and had "a feeling" that there had to be some relationship between the two short sides and the long side. Not long afterward the child learned about Pythagoras' Theorem stating $c^2 = a^2 + b^2$. What was a "gut feeling" before became a clear and precise spoken statement afterwards that could be evaluated and discussed (even mathematically, which is a much more precise "language" than spoken language). The step from taking the internal description of relations between triangle sides to a statement

---

[20] This is also true for aural information (adjacent frequencies are coded by adjacent neurons) and somatosensoric information (adjacent locations on the skin are coded by adjacent neurons) [Gol02].

(which is even a mathematical statement and therefore even more precise) that could then be transported to other people easily. Still, the key remains in the internal description, in the "feeling" that there is something that correlates (see also section 1.4).

There exists a "language", which is outside the scope of the spoken or written language that we use when communicating with other persons. Our inner world has means to describe facts, events and relationships by using a description that is usually not available to conscious contemplation[21].

1.7 Consciousness

Consciousness can be described as continuously firing neurons that follow certain patterns. Although true, this description is as useful as describing the quality of life in a city by listing its telephone book. Consciousness is the most complex function of the human brain. Damasio separates consciousness into *core consciousness* and *extended consciousness* [Dam99], where core consciousness refers to abilities that humans share with, for example, some vertebrates, while extended consciousness is at the current state of research unique to humans and refers to what is commonly called "consciousness". Core consciousness gives an individual a "self-sense" in the current here and now, without any history. It does not allow predictions of the future and considers the past only as far as the events that recently happened [Dam99]. Consciousness is a private phenomenon that cannot be shared and remains restricted to the person experiencing it.

Seen from an evolutionary standpoint consciousness contributed to the survival of an individual. The ability to reflect upon what is happening, to get a "feeling of what happens" [Dam99] allowed an individual to better adapt to its environment and react in a better way.

At the current state of technology, a system that implements extended consciousness, is not expected to occur soon. Assuming that consciousness strongly depends on all other steps of development human beings have gone through (including development of all different brain areas), the closest goal that seems achievable is the implementation of perceptive awareness (see [Tam03] and [Pra05b]).

2 Brain Mechanisms and Technical Counterparts

Collecting and processing big amounts of data and extracting relevant information is the base to build a representation of the world. The brain is – seen from the anatomical point of view – an organ like the stomach or the liver. As it is a part of the human body, it consists of cells – the *neurons* – that have their own metabolism,

---

[21] Of course it is possible to reflect on a subset of these internal facts, events and relationships and come up with a description that is "pronounceable" (as in the above example), but nevertheless a lot of what we use to describe our daily life remains internal and is not consciously available to us.

which needs to be supplied with oxygen and nutrients by the body's life support system. A neuron consists of three main parts [Sol02]: the cell body or soma, which is the central part of the neuron, the dendrites and the axon. While there is only one axon per neuron (which can split into several branches), there are multiple dendrites, which enable the brain to have a vast amount of interconnected neurons – the basis for the complexity of the brain. Axons are responsible for sending out information from a neuron; dendrites provide input into the neuron. Axons and dendrites of different neurons connect with each other over a gap called synapse, where the information transport between one neuron and another takes place on chemical basis. Some types of neurons can have thousand and more dendrites, thus allowing communication between lots of neighboring cells. In total, the human brain consists of some billion neurons with some magnitudes more of interconnections between them.

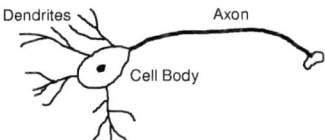

Figure 2.1 A Neuron with cell body, axon and dendrites

Just as the transistor is the basic building block for all modern computer architectures, the neuron is the basic building block of the brain. During evolution this type of cell evolved, which enables vertebrates like human beings to process the information received from the outer world, to learn the regularities in the physical world and to interact with it accordingly. The scientific community puts a lot of effort into understanding the operation of the neuron and the complex network that they build. Recent efforts attempt to simulate neocortical columns, which consist of approximately 60.000 neurons on cellular level by using large-scale computer models and simulating it on the Blue Gene supercomputer [Blue04]. The outcome of these simulations is expected to further contribute to the understanding of the human brain and be used for future simulations with even more neurons.

On the other side neuroscience and brain research have made advances in understanding the structure of the human brain and was able to identify various different brain regions as being responsible for different functionalities. A lot of different theories and model evolved from the attempt to understand the human brain. Early model tried to assign specific brain regions to certain abilities, for example the brain model of Franz Gall in the early 19th century. He identified "organs" in the brain, which were responsible for the sense of orientation, colors or sound as well as organs for theft, jokes or art [Lur01]. It soon turned out that this mapping was too ambitious and overshot the mark of functional brain mapping (see also section 1.1).

Today it is understood that simple physiological functions can indeed be localized in certain brain regions (such as the senses or motion), but higher mental functions cannot be mapped onto one specific brain region, but rather involve many different parts of the brain [Lur01].

Analyzing the human brain in this way is a difficult task and contains potential for fallacies. Assume one would have to analyze the functionality of a car without having access to it directly, without being allowed to open it, take it apart or measure something inside the car. Mere observations on the outside are allowed and even those are limited. We restrict ourselves to observing a point on a screen that shows the current position of a car. A correctly working car could thus be observed as a device that is able to drive at a constant speed in straight direction (as one of the many possible observations) – shown as a straight line on our screen. Now we assume that the front left tire blows and the car is forced to drive only on the metal rim on the front left side. The observation from the outside would notice a fault in the front left area of the car and show a car that cannot drive straight any more, but runs around in a circle around the left front. The next observation would be done on a car with a faulty front right tire, which again would force the car to go around in circles, only this time to the right side. Taking all three observations together – without a fault the car goes straight, with fault 1 the car makes left circles, with fault 2 the car makes right circles – we could deduce that the two areas of the car in the front left, or front right, respectively, are responsible for controlling the ability of the car to go straight. Moreover, they both need to be in balance, because, if one of them fails, the other one is dominant and forces the car "away" from its side.

Describing the front tires as "keeping the car in straight direction" devices for a car is a description that lacks the engineering insight into the functionality of a car, but is reasonable when observations can only be made from the outside. To fully understand the function of tires one has to do a lot more observations and combine them in the right way. Unfortunately, the brain is much more complex than a car. To better understand this complexity, we take a look at an example that is a bit closer to the complexity of the brain: a microprocessor.

In autumn 1994 a bug in the recent Intel Pentium processor became public: under certain circumstances the processor made mistakes in simple arithmetic calculations. Most of the time these errors were small and negligible, but sometime the result was seriously different from the expected result. For example, the calculation

$$z = x - \frac{x}{y} * y$$

should always evaluate to $z = 0$. However, with $x = 4195835$ and $y = 3145727$ the faulty processor produced $z = 256$ [Pet97]. The reason for this error was found to be a lookup-table in the processor that is used to speed up calculations; this table contained 1066 values, but five of them were not correctly accessible. If a calculation required one of these five elements, it produced incorrect results.

How was it possible to find this bug? A processor like the Intel Pentium contains a few million transistors. The functionality of a transistor is fully understood, as well as the functionality of all other structures that can be found in a processor. Theoretically, it would be possible to identify the "faulty transistors" (i.e. the structures that contain the five faulty table entries), determine their position and isolate the error to be located at these transistors. Maybe one could even correct the error in an existing chip by changing structures or connections[22] and thus remove the fault.

Reality in information technology shows, however, that an error like the one described above could never have been found, if the processor design were available only as the final silicone chip containing transistors and connections without any additional information. Instead, the processor designers have used different models to create a working processor of this complexity: a model on the lower layers describes logic functions by a set of *logic gates* – a logic AND or OR, for example. Gates are groups of transistor and the description of their functionality is more abstract than that of a transistor. Other elements can store information (called *flip-flops*) and also consist of a number of transistors. At this level it is, for example, easy to build a device that is able to add two numbers and store the result. Using logic gates and flip-flops as basic building blocks the next level of abstraction is reached by using another model that describes *sequential logic systems*. Such a system maintains some kind of state and changes this state according to information it either receives from the outside or generates by itself. This way a system can be in a specific state, wait for some input and then go through a sequence of actions to produce a result. At this level, it is possible to describe simple machines like a coffeemaker that waits for coins to be inserted and then delivers hot coffee to the user.

Based on a sequential logic system the next step is to create a more flexible system that is not designed for one specific purpose, but is able to fulfill different tasks. This is done by defining *instructions* that the system is able to execute (e.g. "add these two numbers" or "if this number is equal to zero, stop execution") and let it execute these instructions one after the other. The order of instructions and the length of the sequences is arbitrary and thus allows to build a *programmable system*, which is able to fulfill a broad variety of different tasks. Using this model it is possible to build, for example, a pocket calculator, which is able to process numbers entered by a user and apply basic arithmetic like addition, multiplication, or square root, according to the requests of the user.

The list of models does not end here, it can be extended a lot further. Every time a new model is introduced, it builds upon the model below and the abstractions that

---

[22] Corrections like these are actually done in processors today, only on a different level. Aside of the static silicone – the hardware – a processor also contains *microcode*; microcode describes how the processor shall use its resources when executing an instruction. If a bug in an already manufactured processor is found, it is sometime possible to correct it by modifying the microcode of the processor.

have been done there. Designing a coffee machine by using only the model for transistors is an error-prone, almost impossible task. However, using the right model allows to abstract and simplify unnecessary effects and focus on the relevant facts, making coffee machine design a practicable task.

The big advantage of information technology over neuroscience is that all of the components that are used have been designed by humans. Thus, we know every detail about every single component and can use this knowledge to create models that are appropriate, that focus on relevant parts and leave out unimportant parts. Although it is still a challenge to create a model that is comprehensible and easy to use (in the short history of computer science there were always a few scientists who pushed things forward), it is possible to find such a model and – more importantly – prove that it is correct.

When trying to apply this methodology to neuroscience there are a few problems that we face: a processor today consists of some 100 million transistors; the human brain, on the other hand contains some billion neurons, which are connected by about a trillion dendrites [Sol02]. The complexity of the human brain is therefore some magnitudes higher than the state of the art in computer science. Next, the human brains is a biologically grown organ; although in general every brain works the same way, each individual brain has its own layout; even worse, the connections between neurons change over time, which reflects the individuals personal experiences. Processors are, on the other hand, designed (using all of the above described models) before they are manufactured, and are then produced in big numbers, where each processor is (almost) identical to all others. Moreover, they do not change over their lifetime when executing different applications.

The circuit layout of a processor is completely known and understood by engineers. Using the different models, the purpose of each single transistor can be traced up to the highest level and be explained there. The "circuit layout" of the human brain is today by far not fully understood. Using the methods described earlier, a lot has been learned about the functionality of the brain; but compared to the design of a processor this knowledge is still vague and imprecise, especially when looking at higher cognitive functions. And even with the functionality of a single neuron and its interactions with other neurons being available, this is as good as having a good transistor model. We need higher-level models, which abstract groups of neurons to functional units, and group these functional units to even bigger units. This kind of research is currently going on and has achieved great results already, but is still far from being able to explain all functions of the brain (e.g. the Blue Brain Project [Blue04] mentioned above). Single brain regions have been identified and assigned different functions, trying to provide mid-level models for some functional units. On the other side of the spectrum, psychology attempts to understand human perception, human memory, language, motion, and actions. Even higher up in complexity psychoanalysis deals with human beings in their full complexity, trying to understand the highest-level functions like consciousness and the Ego, but of course being unable to build upon models that root in the understanding of the organic brain.

When trying to explain human behavior, it is currently not possible to build a complete model. Too much is still unknown, therefore models have to be designed based on what is known today, carefully making assumptions where information is missing. This is an intrinsic problem of any science that escapes the approach of natural sciences. In order to experimentally prove a theory, natural science requires the experiment to be set up in an environment that guarantees to create reproducible results. Only if an experiment can be reproduced, it is considered, otherwise it is discarded. The approach works, if it is possible to lock down almost all variables and observe only a few or – preferably – one variable. Exactly these "laboratory conditions" make the natural science approach very hard to apply to psychoanalysis (or humanities in general). The state of a human mind is a multidimensional set of parameters, which depend on and influence each other. Thus the reproducible laboratory experiment is inadequate to prove, for example, Freud's theory about consciousness.

2.1 Embodied Intelligence applied to a Building

When applying principles of classical AI and especially embodied and situated intelligence to building automation the problem of a missing body arises. A building automation system is not an insulated system with well-defined boundaries of its body. Instead, it has sensor and actuators in every room of a building. One could suggest defining the whole building to be the body of the system. Although an intuitive suggestion, it is not reasonable, since the system is not strongly influenced by modifications or damage to the building – after all it is dead matter with no vital connection to the system. If a wall is modified by building a new door into it, the system will not perceive it anyhow different as if its environment has changed.

Still the requirement of embodiment remains: what if embodiment is really the base for intelligence? We need to take a closer look at this requirement. What we see is twofold: first there is a very tight connection between how an entity perceives its environment and its sensory input: if a being has eyes, it is able to perceive light within a certain wavelength range. If it has ears, it can perceive aural information of its environment. Somatic sensors provide input about its close vicinity and so on. Depending on where these sensors are located on the entity and how they operate, the entity will perceive its environment differently (an eye located on the front will influence behavior differently than an eye located on the back).

The second issue in embodiment deals with actuators. What has been said for sensors is also true for actuators: depending on the ability and location of the actuators (in fact, their functionality), the entity will interact differently with its environment: everyone who ever broke an arm will agree that the modern world is a lot easier to use with two hands.

Embodied intelligence is based on the fact that sensors and actuators are tightly linked to each other – in the same body – and they have a defined relationship. Aside of the actual sensor information, the information about the sensor itself and its whereabouts is important for further processing. Combined with the fact that learning on this physical level works best when simultaneousness is given – that is, when an action causes an immediate response, we have all ingredients to extend embodied intelligence to a disembodied building automation system: what is needed are sensors and actuators that have a defined spatial relationship. The system has to know where its sensors and actuators are located. Then we need to provide the possibility for fast feedback: actions caused by actuators need to be registered by sensors and processed quickly. This feedback-loop of actions causing reactions builds the base for learning. What we have created is a system that, although it cannot easily be surrounded by an envelope that would define the boundaries of its body, can still be analyzed by means of embodied intelligence.

2.2 The Frame Problem in Artificial Intelligence

The *frame problem* emerged from logic-based Artificial Intelligence. It describes the question that arises when describing facts and events as they happen in the real world, in a formal logic. Commonly one would describe an action and its consequences on the world. A desktop lamp can do two things: it can be turned on and off and it can be moved around. The lamp will change its state when it is switched from off to on. What common sense tells us is that the lamp is now turned on, but the position remains unchanged. If the above statements were however written out formally, they would only allow concluding that the lamp is turned on, but its position would no longer be clear, since it cannot be ruled out that the action of turning on the lamp also has influence on another property of the lamp, the position.

What the frame problem (in its technical form, described in [McC69], a reinterpretation by Fodor can be found in [Fod88]) shows, is that it is necessary to state a frame of reference. Aside of describing, which properties an action affects, it is also necessary to describe, which properties it does not affect (see also [Sha04]) – and this is usually the majority. Formal logic has to deal with the problem to describe what human "common sense" is: ruling out a wide range of unlikely consequences of an action. In order to fit logically, contradictions have to be prevented (in the example above, pushing the button of the lamp too hard may actually move the lamp). Most actions have only limited impact, but some can have significant impact on a lot of properties[23], which brings the question to defining a context, in which the above said is true; only by choosing objects and properties appropriate, it is possible to apply "common sense". In general, it appears challenging to predict every possible outcome of an action in the real world – formally covering every possible case will result in the conclusion that "everything is possible". Even more, the objects that formal logic can operate on need to be acquired in the first place, which requires choosing carefully the definitions (What makes a lamp a lamp?). Logic-based artificial intelligence operates on clearly defined objects (see [Russel03] for a comprehensive description of classical AI), but the acquisition of objects and the concepts behind them is an important issue that is covered in section 3.

3 Symbolization

The word "symbol" is used in many different disciplines, thus its usage might be a bit overstrained. Encyclopædia Britannica [Enc04] defines the term as

> *A communication element intended to simply represent or stand for a complex of person, object, group, or idea. Symbols may be presented graphically, as in the cross for Christianity the red cross or crescent for the life-preserving agencies of Christian and Islamic countries; representationally, as in the human figures Marianne, John Bull, and Uncle Sam standing for France, England, and the United States respectively; they may involve letters, as in K for the chemical element potassium; or they may be assigned arbitrarily, as in the mathematical symbol ∞ for infinity or the symbol $ for dollar.*

Symbolic representation of events, objects, and knowledge in general is a central concept of this work. A question that arises is the connection between symbols and the real world. Harnad in [Har90] asks:

> *How can the semantic interpretation of a formal symbol system be made intrinsic to the system, rather than just parasitic on the meanings on our heads?*

If a symbolic alphabet is defined (together with appropriate semantics and syntax), how can these symbols ever really represent anything else but a combination of other symbols? In [Sea80] Searle explains the *Chinese Room* Experiment: A person, who is unable to speak, read or understand Chinese is sitting in a room and receives Chinese symbols from the outside together with a set of rules. While the symbols are incomprehensible to the person, the rules are in English (which the person is able to understand) and describe, how to operate on incoming Chinese symbols to return other Chinese symbols as an output. People on the outside of the room have called the incoming symbols a script, a story, and questions (which is unknown to the person inside). By operating on the incoming symbols and returning Chinese symbols using

---

[23] In a thought experiment, Fodor creates the concept of a *fridgeon* [Fod88], which applies to any existing particle in the universe, if and only if Fodor's fridge is turned on. Obviously, the action of turning Fodor's fridge off has significant impact on a lot of particles.

the rules, an observer from the outside may come to the conclusion that the person inside (or the system as a whole) is able to understand Chinese, because the system responds to Chinese questions with Chinese answers. Even more, if the system is left unchanged, only the Chinese Symbols are replaced by a story and questions in English, the system operates just as well. Although the person in the room now actually understands what is written, the outside observer may come to the conclusion that the result of the Chinese version is as good as the English version.

The point of interest here is how to find a connection between the symbols of a system that operates symbolically (or a mind, which is often modeled as such) and what the creator of the system has intended as the meaning of the symbols. Or, to apply it to the system envisioned in this work: how can the symbols be connected to the sensor data that the systems uses as the interface to the real world? To build such a system, the intended meaning of symbols needs to be replaced by links to real world information. Otherwise, symbols would be combined with other symbols to generate even other symbols and we face another problem, sketched by Harnad [Har90]: If a person wants to learn Chinese, but all that is available is a Chinese-Chinese dictionary, the person is never able to learn the language, because one meaningless symbol would always be explained by a set of other meaningless symbols. The symbol alphabet is closed in itself without any way to obtain understanding.

In [Yu04] a multimodal interface is introduced that allows a machine to learn words of human language and assign them to everyday tasks. The system learns unsupervised; test persons are equipped with an eye tracker, a camera, a microphone and position sensors (for head and arms) that allow the system to have a first-person impression of the tasks that are done by the test persons. The data from the sensors is used in combination with a spoken description that the test person gives while fulfilling a task. Because the meaning of the spoken description is also contained in the sensor data, this redundancy is used to create a combination between the utterances and the sensor data.

This is an interesting example on how to provide symbol grounding. If we call the associated utterances "symbols", then a set of symbols is obtained, which is intrinsically grounded to sensor input. The semantics of these symbols is already given and the phoneme strings that label the symbol have a high probability of sounding similar to the actual word that a human would use to describe an object or activity.

Harnad criticizes symbolic systems for exactly this lack of grounding: if symbols (or "tokens", as they are also called) are manipulated by explicit rules and the rules are purely syntactic, there are not intrinsic semantics in the system. Although the system may be semantically interpretable by a benevolent observer, the meaning in the symbols is merely "parasitic" [Har90], as has been shown in the Chinese Room experiment. He proposes to use a connectionism (i.e. neural networks) system for these symbols to be grounded; [Yu04] uses more advanced methods to achieve this grounding, but the idea remains the same.

The ARS system operates on a similar basis: grounded symbols (called *microsymbols*, see section 4.1) are the lowest layer of symbolization and at the same time provide the link between sensor data and symbolic processing. Starting from microsymbols, the system is able to operate solely on symbols. Since a basic, grounded symbol alphabet is defined, new symbols can be created on top of it – and they will inherit the grounded meaning of their "parent"-symbols.[24]

Choosing the symbolic alphabet has to be done carefully, because the selection of the alphabet may influence the ability to find a solution to a problem. Examples are the Roman numerals that are inferior to the digit-oriented Arabic numerals for arithmetic operations. While a hard-crafted symbol alphabet – like the one used in the ARS system – can be tuned to avoid unsolvable problems, this may be of interest for future versions of the system, where the symbols are obtained automatically.

---

[24] Harnad states an example: if the symbol "horse" and the symbol "stripes" is grounded, it is possible to create a symbol "zebra" by defining it as "horse and stripes" [Har90].

PART II: System Design

This part of the thesis contains the descriptions of the system and its design by taking the principles and mechanisms described in Part I and applying them to applications in the domain of building automation – the ARS system. First, the system model is described in chapter 4; here symbolization and its different levels is covered, and perception as it has been described in Part I is mapped onto the system. Chapter 5 contains descriptions of the sensors that are used to achieve the goals of the reference implementations. The sensor behavior is abstracted in a way that makes them generally usable for the system and the models used for the sensors do not depend on the specific properties of the sensor. The system operates with these generic models to create symbols and the world representation.

Beside the information that is gathered by the sensors which provide the interface to the outside world, the system also relies on predefined knowledge that it can use to augment the information of sensory input. This is described in chapter 5.7.3, where also the issue of information storage is briefly covered: while the memory available to the human mind is complex and not yet completely understood, the task of the system to store what information has been gathered and what actions have been triggered can be achieved more easily by storing the according symbols.

The subsequent chapters describe the whole system design starting from the primary source of information – the perception – and following the system until it triggers actions. Chapter 6.2 describes what is needed to implement perception, the basic requirement for the system to perceive the world. Chapter 8 builds on perception and shows how a world representation is created based on the updates that permanently come from perception. Here the difference between the inner and outer world are also covered. Chapter 8.4 explains how the world representation is used to gather scenarios; these scenarios are then used in chapter 8.5 to let the system react accordingly – which is the task of the Action Subsystem.

Chapter 9 contains the description of a set of reference applications that are used to show the functionality of the system by using the mechanisms described in the previous chapters. Since the intended use of the ARS system envisions surveillance tasks as well as care and supervision of different groups of persons (seniors or children, respectively), both the basic tasks like person tracking and object tracking and the sophisticated applications of, for example, geriatric care are covered.

Since the system does not implement any facilities to learn new facts or rules during normal operation, there are also no statistical methods included here. The area of machine learning by statistical methods is part of the project that this work is embedded in [Sal05], but is left out in this thesis.

Part II defines the requirements that the implementation in Part III needs to fulfill. The concepts and models described here are general, while the implementation in the next part focuses on the reference applications only. In this part the foundations are laid by constructing models that build the base for the implementation of the system – in a way, it is a "duty book" for the system, describing the functionality of the parts that make up the system. The bridge between information technology and the biological, neurological and psychoanalytical sections of this thesis is built here: bringing together what has been said before and merging the previously described disciplines and mechanisms into one real system. However, implementation is not described here. Nothing is said about which tools from information technology is actually used to implement the functionality. At the end of this part, the reader has a description of the intended functionality, which should be comprehensible by reading through the previous text.

4 Symbolic Processing

Applications in today's building automation systems use fieldbus systems (e.g. LonWorks communicating over the LonTalk protocol [Ech94]) to connect a number of sensors and actuators. The sensors are the interface to the physical world and provide the information necessary for the system to operate. In classical building automation, it is common to have one sensor for each task for reasons of costs and efficiency (see also [Pal03]): room temperature is measured by one sensor per room and presence detection is achieved by installing motion detectors, each with a separate area of visibility. What we see is a one-to-one relation between a sensor and its intended application. Historically, automation technology has therefore focused on single sensor values and designed applications to handle these values. The advantage of this approach is that sensory input can be modeled in great detail, using well-known and deterministic mathematical models. The disadvantage is the limited view a system can gain from the world surrounding it and the strong dependence on the correct operation of every single sensor.

With the increasing number of sensors installed in a building or factory, this relation is extended in two ways: one sensor is used for more than one application and applications attempt to use more than one sensor to gain information from the physical world [Lon02]. This development required to rethink a paradigm: applications cannot be designed to process sensor data; there is a need to introduce layers that separate sensors from applications. *Sensor fusion* is an approach that implements this concept. In [Elm02] sensor fusion is defined in the following way:

> *Sensor fusion is the combination of sensory data or data derived from sensory data in order to produce enhanced data in form of an internal representation of the process environment.*

Instead of using single sensor values, an application deals with pieces of information that represent the status of the physical world. These pieces of information can originate from a sensor, an aggregated set of identical redundant sensors or can be created by using different types of sensors that are merged together. A sensor value from a motion detector, a light barrier and a tactile sensor in the floor can therefore be combined into a valuable piece of information, namely "there is movement at this position".

Sensor information is condensed and the amount of data is reduced, while the quality of information is increased. Instead of processing all incoming sensor information in one application, a hierarchical structure reduces the amount of data, but improves the quality of information. Throughout this work, such a piece of information is referred to as a *symbol*. Except for the very lowest layer, the perception module, all processing of information is based on symbols, not on sensor values.

The author is aware that the term symbol is widely used in various different domains and is therefore bound to be misunderstood. Symbols are visual or linguistic signs that refer to something – e.g. a fact, a piece of information or an event. There are religious symbols as well as traffic signs or mathematical symbols. Symbols in data communication are used to describe completely different concepts than the symbols a psychoanalyst uses to describe and interpret a patients' dream. Because of this ambiguity of the term symbol, a description of how this term is used here is given below, followed by further information in the following sections and a description of the implementation of symbols in chapter 11.

A symbol is a piece of information on which the system can perform operations, modify it, create and destroy it. Symbols contain information originating from the physical world as well as additional information that define their meaning. The symbols used in this work are collections of information, which are contained in the symbol itself and in its *properties*. A symbol can have one or more properties (e.g. the position and the current activity of a person) that are part of the symbol and cannot be separated from it. Properties can be updated frequently (e.g. if the position of a person changes) or remain unchanged.

The complete description of the implementation of symbols and properties is covered in chapter 11. The following sections explain how symbols are used in this work as the basic building blocks for processing information.

4.1 Symbolic Model

The model that is used here is taken from the work of neuroscience, psychology and psychoanalysis (see also [Bra04], [Die02], [Die04], [Die04b]). The attempt to apply the models that are known in these fields to the area of building automation yielded the hierarchical model shown in Figure 4.1. There are three layers of symbols: *microsymbols*, *snapshot symbols* and *representation symbols*[25]. All of these exist at

---

[25] The model is extended further in section 4.1.3, where more types of symbols are introduced. However, to bring forward the idea of symbolic data processing, these three types of symbol will suffice.

the same time, but on different levels (a fact that becomes important later in the text). Symbols can be created, their properties can be updated, and they can be destroyed. Symbols are shown as cuboids of different volume in Figure 4.1, indicating that their level of sophistication grows with each level. Also, the number of symbols is different on each layer: while there are a lot of (small) microsymbols on the lowest layer, there are only a few symbols on representation layer, each of them representing a lot more information (or information of higher quality).

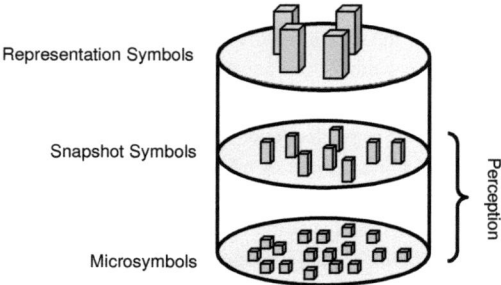

Figure 4.1 Levels of symbolization

The three types of symbols shown in Figure 4.1 are defined as follows: a *microsymbol* is a symbol that is created only out of sensory input. Similar to the many different sensations that the human brain has to process every second, a microsymbol is created out of a few single sensor inputs at a specific instant. Microsymbols are created every time the real world has changed and this change has caused sensors to trigger. There are not many interrelations between microsymbols, they are rather orthogonal. Microsymbols are symbols with event-like character, meaning that they exist in one instant (or a very short period of time, respectively, see section 12.5.1). The step from sensor values to microsymbols includes a first processing of the signal in terms of debouncing and discretization. Sensors that are able to create continuous data streams are treated specially in this lower layer, but also result in producing microsymbols (see section 5.6 and 5.7).

A group of microsymbols is used as the base to create a *snapshot symbol*. These symbols represent a part of the world at one point of time (or a short time span, respectively). All snapshot symbols taken together represent how the system perceives the world in one instant. If available microsymbols are, for example, footprints on the floor, interrupted light barriers and a face that has been recognized by a camera, the according snapshot symbol that is created out of these microsymbols is a person, present at the current time at the current position. The creation of snapshot symbols implies a first layer of associations between microsymbols. Since all symbols are "hand-crafted", these associations are designed manually for each symbol. In a future step, the model will be extended to gain the ability to create symbols by itself; this ability to learn symbols is, however, left aside here.

The next level of symbolization is the representation of the world (see also section 1.2). *Representation symbols* are – compared to the previously described symbols – only few in number and are hardly ever created or destroyed (only their properties are updated regularly). On representation level, the system has information about the current state of the world together with the history of what has happened. Based on snapshot symbols the system uses all momentarily present perceptions to create a consistent and continuous representation of the world. Representation symbols are the first level, which applications use to infer about the world. The snapshot and microsymbol layer below shall not be used by applications, since applications represent higher cognitive functions, which shall operate on the according set of symbols[26]. It is understood that the mind does not operate on, for example, single impressions on the retina, but rather on the perceived image[27]. Therefore, the applications described in chapter 9 use the representation to find scenarios instead of analyzing single snapshot or microsymbols.

4.1.1 Symbol Hierarchy

The creation of snapshot and representation symbol is shown in Figure 4.2. A set of microsymbols triggers the creation of snapshot symbols; new snapshot symbols again cause an update of the representation symbols. These two levels belong to the perception of the system (indicated on the right of Figure 4.2). What is not shown here is the continuity of representation symbols: while microsymbols and snapshot symbols operate in time instances, the representation symbols operate in time periods and thus have a long lifetime. Typically, a representation symbol is used for one specific person that exists in the world. While this person causes the creation of a lot of microsymbols and snapshot symbols (e.g. when walking from one room to the next), there is always just one representation symbol for the person; this representation symbol is permanently updated by the information coming from snapshot symbols.

---

[26] Implementation does, however, not prohibit applications to access microsymbols or snapshot symbols.

[27] Although the human mind is able to focus its consciousness on one specific detail – which may be translated into one specific microsymbol in this system – this is not the common case. The higher cognitive functions operate on facts like "the person has left the room" rather than "the latest footprint is not in the same room as the previous one" without being aware that both footprints belong to the same person.

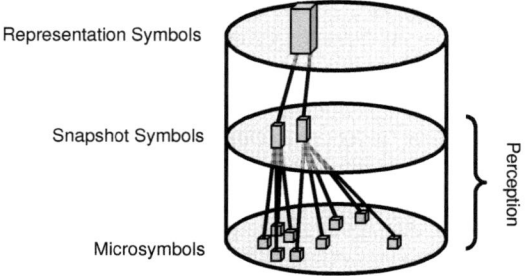

Figure 4.2 Creation of symbols

The model that is shown here is strictly hierarchical, meaning that symbols on one layer may only use information available on the adjacent layer below. It is not permitted to use, for example, microsymbols to create representation symbols. The implication of this hierarchy is that a symbol on one level needs to contain all the information from the layer below, which might be needed on a higher layer. Compared to the classical automation pyramid (see, for example, [Pfe87]) the symbolic model is similarly hierarchical and can be put aside of the data processing in field level, process level and cell level in factory automation. The highest level of the automation pyramid, which employs global networks on company level, has no counterpart in the symbolic model, since cooperation between systems in different buildings is not intended.

The human brain does not follow this hierarchical model as described above, except for the evolutionary oldest parts of the brain. Thus, a design that attempts to consider the architecture of the human brain would have to aim at a maximum of interaction between all levels and not restrict interaction by a hierarchical model. On the level of symbolization, however, that is described here, it is nevertheless sensible to apply the hierarchical model, because it allows for a simplified design without sacrificing the ability of cross-communications. Should it become necessary to access low-level information from representation level, additional feed-through symbols can be defined to provide the necessary information on a higher level[28].

4.1.2 Symbolization in the Inner World

The model introduced above is a generic description how sensor input is processed to create a world representation. The system, however, knows two different worlds (or world representations, to be precise): the representations of the inner and the outer world (see chapter 8). While the principles of the process described above hold true for both representations, it is still important to understand that these two are actually

---

[28] This would be similar to a simple "focus of attention", where the system carefully examines a small detail of the world representation (just like a human would examine the sting on the finger after touching a thorny rose).

different representations. Both worlds contain microsymbols, snapshot symbols and representation symbols, but the two representations are only loosely connected.

Especially in the case of the inner world representation, the microsymbols that are the base for other symbols may originate not only from sensor data, but also from data that represents the current (maintenance) state of the system. For example, a broken communication link to one of the sensors may not be detected by a sensor, but rather by a watchdog timer, which expires because an issued request does not receive the expected response in time. In any case, a microsymbol is created and the system operates identical in both worlds (of course with different sets of symbols) to create snapshot symbols and representation symbols).

4.1.3 Augmenting Symbols on Representation Level

The world representation, which is the topmost level shown in Figure 4.1 contains the systems' view of the world (both of the inner or outer world). Since the representation layer is the one where applications operate on, it consists not only of representation symbols, but also of symbols that are created by the applications.

When an application runs, it searches the existing world representation for scenarios that the application knows (e.g., an elderly person has collapsed on the floor). The events that are required for the scenario to take place can all be found on representation level. Therefore, the application augments the representation by noting that it has found a scenario. It does so by creating a *scenario symbol*. This way it is later possible to study the output of applications[29]. Additionally, an application can create higher level scenarios by linking together lower level scenarios of other applications; this way the hierarchy can be even further extended by having lower level applications looking for simple scenarios and higher level applications using these scenarios to find more complex scenarios. One level of hierarchy between applications is already foreseen in the system: application services fulfill different tasks that can be used by other applications. However, application services merely update representation symbols, they do not create scenario symbols.

A scenario symbol represents the information that a scenario has successfully been detected. At the end of a sequence of events, an application creates a scenario symbol. Usually such a recognized scenario requires some action to be taken. Although the ARS system is designed to be mainly passive and help to support human operators or personnel, in some cases it is still necessary to trigger an action.

For this purpose, two types of symbols are defined: *action symbols* and *action series symbols*. In the common case, the system will take an action, which consists of a sequence of single actions[30]. It therefore creates an action series symbol, which is a

---

[29] All symbols are stored in a database and kept for further reference (see section 12 4.2).

[30] The trivial case of an action series symbol consists of only one action, e.g. notification of a human operator.

symbol on representation level. An action series symbol represents what the system needs to do to react properly on a scenario that has been detected. Further processing of action series symbols is done in the action subsystem (described in detail in section 8.5), which is able to process action series symbols and break it down into several action symbols. These symbols are also part of the representation of the world, since they indicate what the system has done to react on a situation and at which time which action has taken place.

4.2 Symbolizing Physical Values

Physical values like a time period or the speed of an object are accessible by common formal methods and are handled by mapping the magnitude of the physical value onto a numeric value (for example, a scalar). Using mathematical operations, the value can be combined with outer values to derive a result. However, this is not the only way to model the physical world. [Gol02] describes perception in the human brain, where specialized neurons detect motion and direction; [Wan92] identified neurons in pigeon brains that are able to detect collisions with moving objects. This research indicates that processing of, for example, motion, time, direction and other physical values are processed differently in brains, starting from an early time in the evolutionary process. Under the assumption that neural representations are the link between symbolic processing and the real world, this indicates that a processing mechanism based on symbolic representation is closer to the way the human mind processes physical values than by using formal mathematical methods.

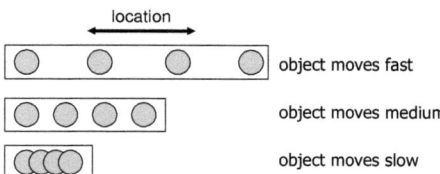

Figure 4.3 Image comparison to symbolize physical time

In a first step, processing of physical values by symbols requires abstracting from an actual object (or person) to a generic object. Figure 4.3 shows an object abstraction for motion; the object moves with a certain speed, which can be symbolized by one of three motion classes. Depending on the motion of the object, one of the symbols is selected to represent the motion of the object. In physical values, the three symbols would represent different speed ranges, but symbolic representation does not rely only on one physical value (e.g. the time difference that is measured in which an object passes two locations).

Similarly, physical relations like time instants relative to the current time have to be modeled symbolically: instead of a time difference measured in time units, the symbolic representations of "before" and "after" are applied. Additionally, the time axis does not necessarily have to be linear: personal perception of time tells us that events that have occurred a long time ago are not stored in the same resolution as recent events. By introducing different resolutions depending on how long ago an event happened, it is possible to model this ability.

Symbolic representations, however, do not depend on only one physical value, but consider different sources of information, like the spatial context in which an object is embedded. In addition, when looking at a specific object, the sound that is produced by that object while moving, contributes to the description of the motion of an object. In case of persons or objects, movement profiles of motion sequence (e.g. running) described as the relative movement of body parts or angular motion descriptions of joints are evaluated to give clues about motion. If it is possible to utilize this additional information in a technical system, motion (and other physical values) can be described symbolically, thus substituting classical measuring methods.

5 Sensory Equipment

The ARS system relies on the input of diverse and redundant sensors. Such sensors are today available and will in future be even more affordable. New technologies allow having not only cable-bound sensors, but also wireless sensors, which have sufficiently long battery lifetime or scavenge energy from the environment [Mah04]. The sensors described here are abstract models that represent the main functionalities of sensor. Specific properties are not taken into consideration, instead a generic description is used wherever possible; if it is necessary to consider specific properties of a sensor (such as the resolution of a camera in section 5.7), it is explicitly mentioned. In further steps, it may become feasible to model sensors to match their specific physical properties and abilities. Since the simulator that is used for the ARS system (see section 12.5) uses the same models, it is appropriate not to differentiate between real-world sensors and simulated sensors. Even more, the system shall be able to perceive its environment correctly using only the abstract model of sensors.

The sensor types described here have been selected, because they contribute to the goals of the reference applications listed in section 9: location information of persons, identity of persons, activities of persons and object recognition. Different sensors have different abilities, with the cameras described in section 5.7 being the most advanced sensors and in addition the only ones that are used for object recognition.

All sensors have a position that indicates where the sensor is mounted. This position is defined by one point in three-dimensional space[31]. Sensors also have a sensitive area, indicating the part of the world they "see".

---

[31] This implies that the size and shape of sensors is not taken into consideration.

5.1 Light Barrier

Light barriers are used in rooms or halls to provide position information. Commonly a light barrier uses infrared light that is focused on a receptor. When the beam is interrupted by an obstacle or a person, the receptor is not illuminated and triggers an event. Another way of implementation is to include light emitter and receptor into one device; this eases installation, since only a passive reflector is needed on the other side. The functionality as required for the ARS system is not affected by this variant; therefore, the schematic drawing in Figure 5.1 separates light emitter from receptor.

Light barriers have a limited maximum distance, since the beam broadens as the distance increases; longer distances require better focusing of the beam, which increases the price of the device. For building automation, a distance of a few meters is sufficient, since light barriers are usually mounted transverse to, for example, a hallway. The light barrier model simplifies a real-world light barrier in two aspects: the broadening of the beam along the distance between sensor and receptor is not relevant and therefore the beam is modeled as a line. Furthermore, there is no latency between the interruption of the light beam and the triggering of the sensor.

The information that can be gained from a light barrier (Figure 5.1) is imprecise position information along the axis of the beam (x-axis) and precise position information in the two axes normal to the beam (y-axis and z-axis). This information is binary with respect to the different output values of the sensor: a light barrier can be either interrupted or not.

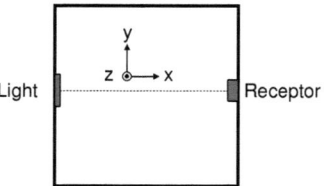

Figure 5.1 Light barrier

Within the ARS system, light barriers are used for two purposes: for once, they contribute to gather position information about persons. For this purpose they are mounted close to the floor in certain distances; secondly, they are used for determining the size of persons. In this case, a set of light barriers is mounted on top of each other, for example in a doorframe. Depending on the gap between two barriers the height of a person can be estimated roughly, which will suffice to contribute to the decision, whether a person is a child. More details can be found in the application descriptions in chapter 9.

5.2 Motion Detector

Motion detectors are triggered by movements within their sensitive area. As opposed to light barriers, a motion detector does not have a single line as sensitive area, but is open to a (usually configurable) angle that it reacts on. If there is movement within this angle, the motion detector is triggered (Figure 5.2). To ensure that only relevant changes in the sensitive area cause the sensor to be triggered and to avoid that the sensor is incorrectly triggered by, for example, thermal noise, there is a threshold in the amount of difference necessary to trigger the sensor. A motion detector has only information about its sensitive area as a whole, since it integrates over this area. Because of the threshold of necessary differences, a motion detector may be unable to detect moving persons or objects under the following circumstances: the object is too small to cause a big enough difference; the person or object moves too slow, which also doesn't cause a trigger of the sensor; or the person or object moves along a path that has a constant distance from the sensor. The latter case is only true if the integration over the angle is linear and can be circumvented by nonlinear sensitivity over the angle. The information provided by a motion detector is binary with respect to the output value: either it has been triggered or not.

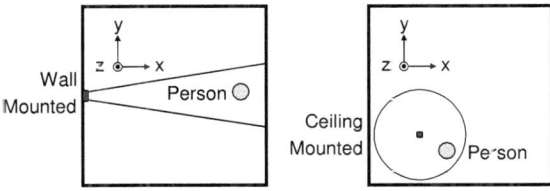

Figure 5.2 Motion detector mounted on wall or ceiling

Motion detectors are used in the ARS system to detect moving persons or objects (if they are big enough, see below) and thus contribute to location information. The model for motion detectors assumes that the threshold for triggering a motion detector sensor meets the requirements for person detection moving at a speed that is within walking speed and running speed. Therefore persons moving too slowly and object that are small (compared to persons) are not detected. A motion detector is usually mounted in a way that the geometric representation of its sensitive area is a complex three dimensional shape (if, for example, mounted close to the ceiling and being tilted so that it faces the ground and the walls). Although deriving the sensitive area of a motion detector that is mounted freely in a room can be done by solving a set of geometric equations (calculations that are nowadays done in standard graphics card adapters in real-time), which will be necessary to completely describe the abilities of motion detectors, the information that can be gained from these sensors does not appear to be worth the modeling and implementational effort that would be required. Therefore, the ARS system simplifies the model by assuming that all motion detectors are mounted parallel to the floor and at a height where they detect persons walking on the floor. This way the sensitive area of a motion detector can be modeled as a triangle as shown in Figure 5.2 with additional height information (z-axis), whereas this height information is the height of the room where the sensor is mounted in. The sensitive area of such a motion detector is then a wedge with a free-form triangle as the base area and a height that is identical to the room height. This wedge is however not the only possible geometric representation of the sensitive area. Other classes of motion detectors can have different sensitive areas, for example a fisheye sensor that is mounted on the ceiling (right picture in Figure 5.2). The adaptation of the model for such a sensor can be done by changing the sensitive area; in the example of the fisheye sensor, it is a cylinder with the sensor in the center of the base circle and the cylinder height being the room height.

Off-the-shelve motion detectors often provide additional functionality: after they have been triggered, their output signal stays active for a (usually configurable) time. This is convenient, if the sensors are used to, for example, directly control the light in a room or hall. For the purposes of the ARS system, this functionality is undesired, since it conceals or at least blurs the location information that the system requires. Therefore, it is assumed that movement detection triggers the sensor output without any time hysteresis.

5.3 Tactile Sensor

Tactile sensors have a lot of different applications in the ARS system. The basic principle is to apply a certain force to the sensor in order to trigger it. They are inexpensive devices and their output information is binary: either a switch is opened or closed. The model for tactile sensors directly reflects this: such a contact can be mounted anywhere in the room; the sensor itself does not have a physical dimension. A simple post-processing process removes mechanical bouncing and provides new values in a time resolution that is feasible for building automation applications. The resulting delay of this post-processing is assumed to be small compared to the required time precision and is therefore not taken into consideration.

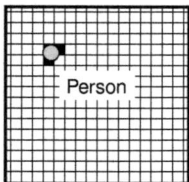

Figure 5.3 Tactile sensors in the floor

The ARS system uses tactile sensors for different types of information: first, they are mounted as an array in the floor of a room or hall to gather information about the location of persons. In Figure 5.3 a person triggers three sensors at the same time, they are shown as black rectangles. Second, switches that are mounted in doors or windows and which are triggered by opening and closing provide valuable information about the current state of the world. A tactile sensor is used in combination with a door or window, in which case it needs a semantic connection to the object that it is triggered by. This object is usually a window or a door, but can also be any other movable object like a drawer, a cupboard door or a simple valve, that can be only fully opened or fully closed. In any case, the sensitive area reflects this object. Door and window contacts provide information about location (person enters a rooms) and activities of persons (person opens a window or door), valve sensors provide feedback for correct actuator operation. Finally, the model for tactile sensors is used to model switches for human users, e.g. light switches. The sensitive area in this case is extended to a rectangular area, which is usually positioned at a wall.

5.4 Temperature and Humidity Sensor

For climate control in buildings, it is necessary to have sensors for the different physical effects that are relevant. The sensors have no physical dimension, since it would not contribute to the applications of the system. The information gathered by the sensors is temperature, humidity or illumination, respectively. The sensitive area is the zero-dimensional point at the location of the sensor. In order to keep the model abstract, additional effects like aging or accuracy are not considered. It is assumed that the sensor is always well calibrated. Depending on the setup, this value can be interpreted differently: if there were, for example, an array of temperature sensors along a wall in a room, the value of one sensor can be modeled as representing the temperature in the area surrounding the sensor, depending on the distance to the next sensors. This interpretation has to be done on a higher level of information processing. The ARS system uses temperature and humidity sensors for HVAC (heating, ventilation, air condition) in order to get feedback about the climate in a room.

Similarly, there are also temperature sensors for a different range of temperatures, namely temperatures that can be dangerous for humans. Such sensors are mounted close to heat sources that are a potential threat to humans, especially children.

5.5 Illumination Sensor

One task of providing comfort to human users is to properly adjust the brightness in a room according to the activity of the user. Therefore, illumination photometers are used to get an overview of the current lighting situation. The sensor is modeled as a small flat area, the normal of which defines the direction in which illumination is measured. Since the area is small compared to the size of the whole room, it can be reduced to a point, where the normal vector is attached. The output is a scalar value.

5.6 Microphone

Microphones are used to detect acoustic signals within a certain frequency range. The ARS system uses microphones to detect sounds and noise that is produced by humans, but does not require high quality sound recording. Since sensor data are stored in a database (see section 12.4), which is infeasible for visual and aural data streams, the acoustic information is not stored as single audio samples there, but instead needs to be preprocessed so that only discrete pieces of information are stored and not the complete stream of audio samples. Microphones can be freely installed anywhere in the room and are used in the ARS system as the following sources of information: a single microphone is used to detect noise in general, therefore a threshold between silence and noise is defined and used to determine, which of the two is currently true. The left picture in Figure 5.4 shows such a situation. The output of this type of microphone is binary and contributes to presence detection of persons and therefore location information. Depending on the specific aural environment, a preprocessing filter has to differentiate between human noises and environmental noises (e.g. air condition, traffic, ventilators).

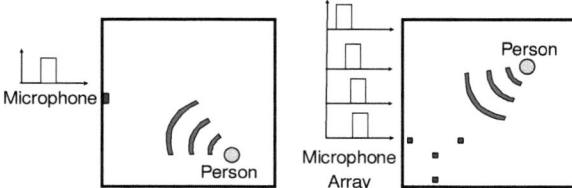

Figure 5.4 Single microphone detecting noise and microphone array detecting location of sound source

The second contribution of microphones is to location information of persons: By using an array of microphones the runtime differences of sound between the different microphones determines the location of the sound source (right picture in Figure 5.4); assuming that a specific sound pattern can be assigned to a human, this again contributes to location information of persons.

For both applications, the same physical devices are used. This way the same device once creates simple binary information – noise or silence, which requires only little effort in implementation, and information of higher quality – location of a sound source, which then requires more sophisticated algorithms. For other applications this principle can be extended and the same microphones that provide location information can, for example, also be used for voice recognition. This multiple usage of the same physical sensor is typical for sensors that provide a great amount of information like microphones or cameras (covered in section 5.7): depending on what is requested from a sensor, the model and preprocessing is adapted accordingly. The output of a single microphone is discrete binary information the output of the microphone array is a three dimensional position of a sound source.

5.7 Camera

Cameras are the most complex sensors that are used in the ARS system and are able to provide a lot of different information. Similar to microphones described above, the system uses cameras as different virtual sensors, which are all based on the same physical devices. A camera can contribute to location information of persons, it is used for identifying activities of persons and it is the only sensor in the system able to identify objects. Cameras for automation (which are used here) are commonly connected using IEEE 1394 (also known as FireWire) [IEEE96]and [IEEE00]), a high-speed serial bus for fast transfer of video and audio data.

5.7.1 Background Subtraction

A single camera that is statically mounted can be used to detect moving objects (or persons) by separating them from the (static) background. [Rad05] contains a comprehensive survey about algorithms for image change detection. The model for this type of camera sensor uses a basic approach: the camera detects differences between consecutive images; if the differences are bigger than a defined threshold it is taken as an image of a moving person (Figure 5.5). Two relevant features are used as an output for this sensor: the number of pixels that have changed are counted and taken as the size of the area and the center of gravity of this area is derived. These two scalar values are the sensor output and written into the database.

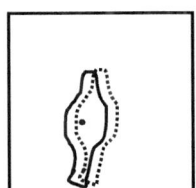

Differential camera image

Figure 5.5 Camera image of moving foreground including center of gravity

Although not a very rich source of information, this virtual camera sensor provides information about movements in a very compact representation, namely two scalar values. Note that there is no depth information included, since a single camera and not a stereo camera is used. It is also important that the size of the area is given in pixels, which makes it dependent on the resolution of the camera, which can be avoided by normalization, e.g. to the visible area of the camera. Other factors are not considered, for example, the zoom of the camera, which strongly influences the size of the two dimensional representation of a real person. Without depth information, the meaning of the area in Figure 5.5 is strongly bound to the camera that provides it.

5.7.2 Face Detection

Not to be confused with face recognition, the task of face detection is to determine, if there is a human face in the current visible area. The output of a face detection algorithm is a feature (which is mapped onto a microsymbol in the ARS system) that has a position and a size. Face detection algorithms are today widely employed, for example as a demonstration application in the OpenCV software platform [Open06]. Based on oriented contrasts in an image, the features of a human face and their spatial relations are used to detect a face. While development of face detection is an ongoing process that attempts to refine reliable detection of faces in different orientations and for partly occluded faces, the existing OpenCV application provides sufficient information to contribute to person detection.

5.7.3 Histograms

The detection of persons is supported by deriving histograms of relevant visual areas. Once a face has been detected, the area is indexed by creating a color histogram. Using this histogram, it is possible to make connections between temporally adjacent face detections and thus contribute to tracking the path of persons. Since the common task of person tracking is limited and does not require identifying persons (or may even prohibit identification for reasons of privacy), the histogram can be extended from the face to the whole body of the person. Naturally, the appearance of a person will strongly vary, depending on the clothing. However, for the duration of a day this can be considered irrelevant.

6 Knowledge and Memory

As shown in section 1.3 knowledge about what to perceive is necessary in order to create a world representation (and to perceive in the first place). Since the system does not learn, all knowledge is a priori available in the system. This includes definition of symbols as well as the necessary facts about the world and its mechanisms. Templates for symbols are used to support perception in recognizing what is known, thus symbols contain knowledge in the way they are defined; the same applies for symbol properties.

6.1 Knowledge about the Environment

Unlike a human being, the system is not able to completely perceive the environment it resides in. Although equipped with cameras, the sensory input is not sufficient to create a complete representation of the layout of rooms, halls and the rest of the building. Additionally, the construction of such a model would be a separate task and is not within the scope of this work. The system needs to have an understanding about the layout of the building it operates in from the very start. This knowledge has to be represented in a way that is understandable for the system.

6.1.1 Environment Layout

Since the system is embedded in the area of building automation and the basic applications require observing persons and their activities, including tracking the position of persons in terms of the room they are in, the model to represent the environment has been chosen to reflect the relationships between rooms, halls and doors. A typical room setup is shown in Figure 6.1. The office environment consists of six rooms that are connected by doors to the hall (or to each other). The central element in the model is a *cuboid*, a three dimensional figure with six rectangulars as its surface. Persons are always located within a cuboid[22], shown as dark gray areas in Figure 6.1. A special form of cuboid is a *door*, which is also has the shape of a cuboid, but a different functionality: just like doors in office buildings, a door object connects cuboids with the special property of separating rooms – as will be shown below. Cuboids are shown in light gray in Figure 6.1. The third type is a *portal*, which is a two dimensional area that defines the border between two cuboids or a cuboid and a door, respectively; these are displayed as dotted lines in Figure 6.1. The convention is such that a person can only change position from one cuboid to the next, if there is a portal between the two cuboids. This way, the system knows, where persons can move and where they cannot (the wall between cuboid 1 and 2, for example, does not contain a portal and therefore cannot be used by persons). Using these three basic shapes it is possible to design an office environment that represents the basic requirements for the system to operate. Based on this model, another layer has to be defined to provide the system with the necessary knowledge about the environment.

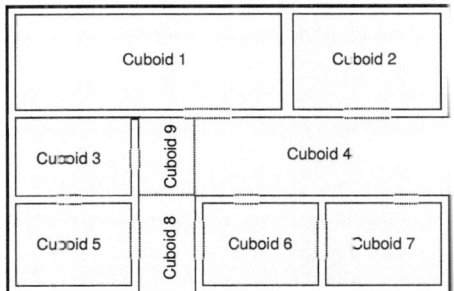

Figure 6.1 Environment model of an office environment

As shown in Figure 6.1 the layout of a real office environment sometimes requires a room to be made up of a set of cuboid (the hall, for example). The definition requires these cuboids to be connected by a portal, if persons are able to move from one

---

[32] In fact, persons are almost always located within a cubicle on the floor, so the three dimensional model could be reduced to a two dimensional model (for one floor); however, the system shall also detect children that climb on tables and similar scenarios, therefore the environment shall not oversimplify the world representation.

cuboid to the next, as is the case for cuboid 4, 8 and 9. Since it is necessary for the system to understand the differences between rooms that are separated by doors (and thus represent separated entities) and rooms that are compiled of a set of cuboids due to their shape, there is another environment model available as shown in Figure 6.2. This model is to be used as an overlay of the previously described model, since it only introduces a layer that defines the setup of real rooms (not cuboids). A room is defined as consisting of an arbitrary number of connected cuboids, limited either by doors or by terminal walls of the building (i.e. walls that mark the outer border of the building). Figure 6.2 shows the rooms of the above layout; most of the cuboids are mapped one-to-one on rooms, only the hall, which previously consisted of three cuboids is now one single room with an L-shaped layout.

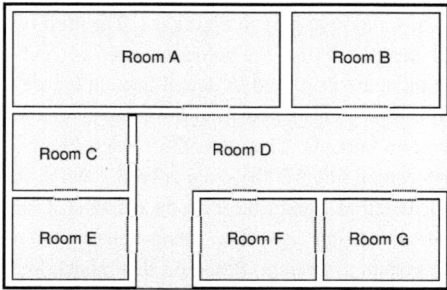

Figure 6.2 Environment model overlay for room definition

The creation of rooms made up of cuboids, doors and portals is done automatically and does not require human interaction. The underlying model allows to determine the borders between rooms by either an outer wall (i.e. a wall, that is at the border of the model) or by a door. For the system, it is important to know, which rooms exist in the building, since it may control the climate for one room and should therefore understand that the hall in Figure 6.2 is a single room rather than three separate rooms. For person tracking it is important to know whether two adjacent sensor values belong to the same person or not. Usually the system can assume that a person triggers sensors in the vicinity of his or her previous position. Then the new information can be assigned to an existing person that is close by. This is however not true, if there is a wall between the two sensors. As said before, the system is not able to perceive the layout of the building. Therefore, it needs knowledge about walls and layout of rooms or, to be more precise, it needs information, whether a person can directly move from one room to the next.

The output of the room creation is the information, which rooms are connected to each other. A graph that contains the connections between room is shown in Figure 6.3. It is important to see that the graph provides a reduced model of the real world: it only represents connections between rooms, but has no information, where, for example, doors in the rooms are located (room F has two doors that lead to room D, which is also not reflected in the graph). The graph shall rather indicate if a person is able to get from one room directly to another.

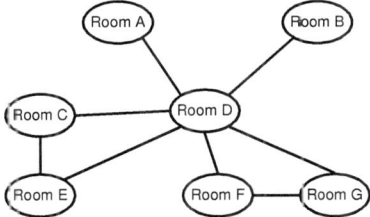

Figure 6.3 Graph with connections between rooms

6.1.2 Static Objects in the Environment

The elementary environment model described in the previous section is the base for all activities in the building. The "world" consists of the static environment, the persons that are moving around in this environment and the objects that are of interest for the system. To get a better representation of the static environment, the model consisting of cuboids, doors and portals has to be augmented by a set of objects that define the layout of the rooms in more detail. Aside of the object that the system knows and is able to perceive, there is a set of objects that are statically built into the environment. If these are relevant for perception, their existence (and their meaning for the system) has to be entered as a priori knowledge. Static objects include, for example, the kitchenette in a kitchen. Parts of this kitchenette may be relevant for the system (like the stove, which is a possibly dangerous object for children), other parts are irrelevant, but still contribute to knowing about possible positions, where persons may be: if there is an object, there can be no person. These static objects are therefore merely treated as "obstacles", which define the room layout in more detail than the cuboids that define the outline of a room.

6.1.3 Hierarchical Structure

An additional layer of modeling the environment is introduced to define a hierarchical model of the building by decomposing the building into entities that consist of other entities. Since this is a purely structural relation, it can be done strictly hierarchical. What is needed is a hierarchy that defines which floor belongs to which building and which room (or hall, respectively) belongs to which floor. To extend this hierarchy, the building is part of a city, which terminates in the root of the hierarchy (Figure 6.4). This way it is possible to integrate different buildings into the same hierarchical structure[33], which is especially relevant when using the same system in different setups, e.g. for testing purposes: by defining in which sub-branch the system operates, a change in setup can be easily done – which is especially interesting for simulated environments as described in section 13.2.

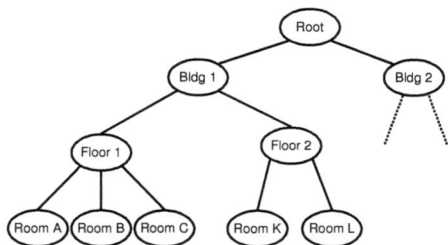

Figure 6.4 Structural hierarchy for the environment

The hierarchy used in this system consists – on the most detailed level – of cuboids, doors and portals; the next level are the rooms as described in section 6.1.1. All rooms on one floor are grouped together into a *floor* and all floors of a building make up a *building*. As said before, the different buildings are grouped in the next higher hierarchy level without introducing any further levels (like towns or countries). This hierarchy – cuboid/room/floor/building – is sufficient for the building automation and observation purposes of the system.

This model of the layout does of course not contain all information about the environment, but it serves the main purposes of person tracking. For more complex perception tasks, further information is added and the model is extended by, for example, modeling that doors can be locked (which requires relations between the locks and the persons possessing the right key).

6.2 Associations

The combination of symbols into new symbols is done using associations. An association takes two or more symbols and creates a new symbol or modifies a property of an existing symbol. A microsymbol that indicates that a person has passed by is used to update the property `position` of the representation level symbol for a person (which does not happen directly, but by using the snapshot level in between). Associations can be of different types, the most prominent one are structural associations like the different representations of room structure shown in section 6.1 or the spatial associations that are used for object recognition (section 7.2.1). Scenario

---

[33] A further separation into countries, counties, etc. is possible, but does not appear to be reasonable here. It would in fact suffice to leave out cities as well and just separate the buildings. This level is however convenient when looking at simulated environments, as they are generated by the simulator.

recognition operates with temporal and causal associations, which define the sequence of events.

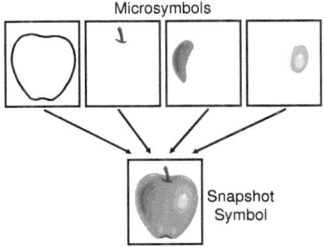

Figure 6.5 Using single features (microsymbols) and their spatial associations to perceive an object

An example of associations for object recognition is shown in Figure 6.5. If an object is to be perceived (the apple figuratively represents an object of interest), the first step is to find different features that the object consists of. When a set of features has been detected and they are in the correct spatial relation, the object is detected. Associations thus can be seen as steps to acknowledge a hypothesis: when the system searches for an object, it builds a hypothesis based on different features that need to be present. New features are associated and thus strengthen the thesis (or weaken it, if they do not match).

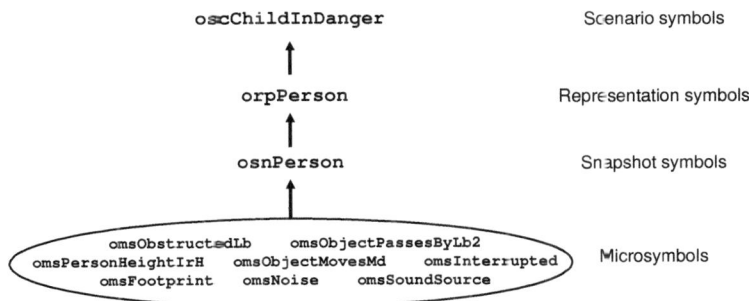

Figure 6.6 Associations between symbols on different levels

Figure 6.6 shows how a scenario symbol is created based on associations of symbols of different levels (the symbols are explained in detail in chapter 11, the example scenario is described in section 9.6). The requirement for the scenario symbol `oscChildInDanger` to be created is that a person, which has the property of being a child is in a room without any other person present in this room. Therefore, the scenario symbol is associated with the person symbol on representation level. To get the representation symbol, it is necessary to associate it with the snapshot symbol, which again depends on different microsymbols collected in the ellipse in Figure 6.6.

7 Perception and Its Prerequisites

The symbols used here are grounded to the real world, as explained in chapter 3; this provides a sound base to implement perception. The system's tasks for perception include determining the location of person, their activities and object recognition. Location of persons is strongly supported by diverse and redundant sensors as described in chapter 5, while detecting human activities and objects strongly relies on the evaluation of optical input provided by cameras.

7.1 Location Information

Many sensors described in section 5 provide information about position of either persons or objects[34]. The concept of position therefore requires a closer look. The different layers of the system have a different understanding of what the position of an object is. A light barrier, for example, can detect when a moving object interrupts its beam of light. This information is used to determine rather precise information along the axes that are normal to the light beam. Along the beam, the light barrier cannot provide exact information about the position, since it does not know where – between light emitter and receptor – it has been interrupted. As shown in Figure 7.1 the position information therefore is imprecise along the beam and exact in the axes normal to it (location information is reduced to two dimension, height is not considered in this figure). The dotted line represents the beam of the light barrier. A motion detector detects changes in the area it observes, which is a wedge, if the sensor is mounted horizontally. A simplified approach models its sensitive area as a wedge, a triangle with a height. Therefore the location information gathered from it is a triangle (or a wedge, respectively, if the third dimension is considered). Tactile sensors that are mounted in the floor and are trigged when stepped on, provide location information in the shape of a point.

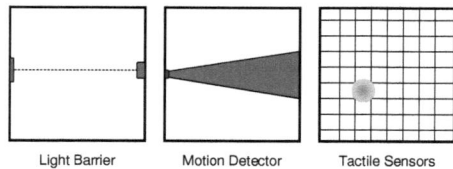

Figure 7.1 Location information gathered from a light barrier, motion detector and tactile sensor

---

[34] In this context, persons are referred to as objects.

At the level of sensor input, it is sufficient to use this type of location information. Contribution to location information by sensors is done through different geometric shapes. For position information of persons in a building (which is the basic requirement for the applications described in chapter 9) a 2.5 dimensional representation is sufficient (two axes for position information on a floor and information about which floor the person is on. For position information of objects, which can be located anywhere in the room all three axes are needed.

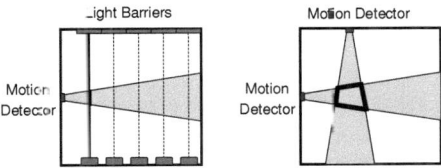

Figure 7.2 Motion detectors combined with light barriers

The ARS system is designed to work with lots of sensor data, which are redundant. In the case of location information, this means that the different geometric shapes that are delivered by the different sensors have to be combined to location information that represents all the contributions. An example is shown in Figure 7.2, where light barriers and motion detectors are used to determine the position of a person. On the left side both motion detector and the leftmost light barrier are triggered at the same time (the triggering of the light barrier is indicated by a continuous line instead of a dashed line). The resulting location information that can be derived is the small line, which is the intersection of the two sensitive areas of motion detector and the triggered light barrier. On the right side, two motion detectors are triggered simultaneously, resulting in location information that is shown by the thick-lined area in the middle.

7.2 Object Recognition

Like the other parts of perception, the object recognition module is based on the symbolization model with its different layers as described in section 4.1. This means that the system perceives objects in a symbolic fashion and the process goes through all symbolic layers, namely microsymbols, snapshot symbols, and representation symbols. The idea behind object recognition is to identify and classify objects and keep track of their position. Some objects are rather static and therefore will not move a lot (maybe not at all), while other objects are being transported, for example, by humans. We know that in the human brain the optical cortex contains a lot of specialized discriminating neurons that react on a certain features of visual perception. On the level of microsymbols, we apply this principle and extract key features of the objects we want to include into our system.

Object recognition is special amongst the other tasks of the system, since it does not use the abundance of different sensors that are available to the system. Instead, it has to rely on the information that is available from the only visual sensors in the system – the cameras. Image processing and computer vision have been researched for a long time already and a lot of knowledge has been gained in this field. This knowledge shall be integrated into the system and not reinvented. Since extracting information out of camera images is a tedious task that requires a lot of research by itself, the results of these efforts shall not be ignored, but rather be used as a base to build symbolic image recognition on top.

7.2.1 Object Feature Description

As an example Figure 7.3 show an abstracted version of a chair as an object that the system shall recognize. The relevant features of a chair can be described as: "A chair needs a flat area to sit on", "A chair has at least one chair leg", and "A chair has a back". On the level of microsymbols, the system needs to be provided with a set of features that identify the object. In the case of the chair, this would be two flat areas that are used as a chair seat or chair back, respectively, and a set of bars used as the chair legs. All these features are objects located in three-dimensional space. Assuming that the microsymbol generator provides the object recognition module with bars and areas, we can continue on snapshot level to check the conditions to find a chair.

The system searches for the objects it knows in the data that are available as microsymbols[35] where there exists a set of flat areas and bars distributed over the whole environment. These microsymbols originate from camera systems that use state of the art algorithms to process camera images and extract edges, areas, objects and other features. The conditions that transform a few of these flat areas and bars into a chair object are size descriptions and spatial relations and other description of features. The chair seat has to be of a size that is feasible for a human to sit on; the same applies to the chair back and the chair legs: they all need boundaries of their size and dimension to qualify for being part of a chair. Spatial relations between chair seat, chair back, and legs are also defined; these can be distinguished in spatial relations between parts of the chair and spatial relations between a part and the surrounding environment. These spatial relations can be described in a pseudo-natural description language, in which the two types of objects are called `flatArea` and `bar`:

---

[35] Note that the system uses hierarchical processing of data: one module has access to the data of the module below, but not further down. Although this definitely does not comply with the way data is processed in the human brain, it is a necessary prerequisite in the software design of the system to get a modular design with communication paths that can be tracked easily; when it is necessary to introduce new communication paths – and thus breaking the hierarchy – the interfaces for this communication have to be thoroughly defined.

Figure 7.3 Characteristic features of a chair

Definition of chairseat
```
orientation of flatArea is parallel to floor
distance to floor is low
```

Definition of chairleg
```
dimension of bar is long and small
orientation of bar is vertical to floor
```

Definition of chairback
```
orientation of flatArea is vertical to floor
```

These are the definitions for parts of a chair. Putting it all together, it is now possible to define the features of a chair:

Definition of chair
```
chair consists of one chairseat
chair consists of one to four chairlegs
dimension of all chairlegs is the same
orientation of all chairlegs is the same
chair consists of chairback
```

Besides defining the basic parts of a chair, there are relations between the parts that have to be fulfilled to keep the parts for further consideration. Chair legs, for example, need to be of same dimension and orientation, otherwise they cannot belong to the same chair. Now we can define the spatial relations between the basic parts:
```
all chairlegs are below chairseat
all chairlegs are connected to chairseat
chairback is above chairseat
edge of chairback is connected to edge of chairseat
```

This description language provides a high-level description of objects using terms and descriptions that are taken from human language and are intentionally kept vague and imprecise. The meaning of these terms has to be described in more detail (i.e. what is "low", "long and small", "the same" for objects in general and a chair in particular) for an implementation to operate with them. Lingual variables like these are known from Fuzzy Logic [Zad65] as well as the fuzzy relationships between the variables.

In the simplified model of the world that the system operates on, only one type of chair is known (the one described above). Obviously, the system will not be able to recognize objects that a human would describe as chairs, but which do not match the description of a chair as the system knows it[36]. One way to deal with this issue is to define more than one description for chairs and create a list of different types of chairs that are treated as different objects (chairs with just one leg and wheels on the bottom, chairs without a back, chairs with armrests, and so on). Still, they share a common property, which is that they belong to the same class of objects.

7.2.2 Anchors

The process of perception is driven by the system: it searches for the list of objects it knows in the current environment. Since objects can be present in many different ways (facing in different directions, partly covered by other objects, the search for an object is done in a sequential fashion, where the system first searches for *anchors* to have a starting point. Anchors are key features of objects that allow the system to build a hypothesis about what it might perceive. In the case of the chair, this anchor would be the chair seat, since it is an elementary property of a chair to provide some flat area, which is parallel to the floor. Using the anchor the system claims one (or more) hypotheses about the class of the object. By evaluating the expressions described in the previous section, the hypothesis is either reinforced and finally becomes confirmed (i.e. the object is perceived) or does not gain enough confirmation to be processed further.

Objects do not necessarily have only one anchor, the recognition can start from different anchors. Instead of using the chair seat as an anchor, the system could also look for any object that is located on the floor and then start evaluating the description to find a chair. In environments where there are a lot of different classes of objects with sometimes similar properties, the description of an object also has to contain negative statements: if, for example, horses are part of the systems perception, a certain description may be appropriate. If, in the next step, zebras are introduced, the definition of a horse has to be extended to clearly specify that a horse does not have stripes.

7.2.3 Object Classification and Identification

When the object recognition module has found an object that qualifies for being a chair (to name an example), it creates a snapshot symbol indicating that there is a chair at a certain position. This information is passed on to the representation module,

---

[36] Especially chairs that have changed their common orientation, e.g. a chair that has fallen over, will not be recognized.

which now has to match this perception with what is known about the world so far. In the case of object recognition this means that the representation module checks, if there were any occurrences of a chair in the vicinity of the current chair snapshot, if the position has changed and needs to be updated or if a new chair appeared in the system. As stated earlier, perception of the system cannot always reliably create a complete state of the outer world, since the current available data might be lacking information. This information needs to be filled in by the representation module.

What has been described so far is the *classification* of an object: what has been a collection of bars and flat areas is combined into a chair. As soon as this process has succeeded once and has found a chair in the environment, this specific chair is no longer just one of many chairs in the class of chairs, but it can be identified (and tracked, as described in section 11.2). Classification is different from *identification* in that first an object needs to be assigned to a category to be able to operate on it. In the next step, the object becomes "individualized" and this requires additional information to be attached to the object. This information is not necessarily related to the requirements for classifying an object, but describes the specific features of one object. In the case of a chair this is, for example, the color of the chair. In the definitions above, the color has not been relevant for classifying an object as a chair. However, once the chair has been found, further information will contribute to better identification – which is especially important on representation level, where objects do not permanently appear and disappear[37], but remain in the inner representation of the outer world, this is the task of Object Tracking, as described in section 9.2. The relevant symbols for Object Recognition are described in the implementation part of Object Tracking in section 11.2, since tracking of objects strongly relies on object recognition.

8 World Representations

The representations that have been introduced in section 1.2 will now build the base for the representations used in the ARS system. The system perceives the surrounding environment and as such creates a representation of the outer world[38]. This representation can only contain objects and facts that the system is able to understand – in case of the ARS system this means everything that the system has available as predefined knowledge about objects, persons, activities of person and the relations between those.

Information processing is done in different modules in the system. First, the system first has to extract information from the sensor level. Based on this data a snapshot of

---

[37] Which might happen on perception level if, for example, a person walks by a chair and therefore obstructs the chair for a moment.

[38] Similarly, it creates the representation of the inner world by supervising its own components and monitoring the health of the internal parts and communication paths.

the current state of the world is created. The information gained in this module is used to create the world representation, which contains not only the current state of the world, but also the history of what has happened. This world representation is the highest level of understanding that the system has of the real, physical world. The system only operates on this representation; anything that is not included in the representation is not available to the system.

Each of these stages is done in a separate module. The perception module uses sensor information to create snapshots, the representation module creates the world representation. Applications that analyze the world operate solely on this world representation. The modules are separated from each other and the interfaces between them are strictly defined. This is a simplification from the understanding of the human brain as we see it today: brain regions are strongly interlinked with each other, allowing cross-links in various different ways. In principle this should also be possible here, however, it would introduce too high a level of complexity. Feedback between modules is therefore only possible using a limited interface.

The representation is the level which is intended to act as an interface for applications and thus for future extensions. Since it is on a higher level than microsymbols and snapshots, the amount of symbols is reduced compared to the lower layers. In fact, a wide field of different symbols from lower layers are collected and used for updating the small set of representation symbols. As an example, we look at the position of a person: whenever a person moves, a lot of sensors are triggered and thus microsymbols are created. The information available from this big amount of microsymbols (with the intermediate step of creating snapshot symbols) is taken to update the property "position" of the representation symbol for the person. While microsymbols are continuously created, the person on representation level is created only once (when a person enters the system) and afterwards continuously updated (the symbol hierarchy in Figure 4.2 also shows these relations).

8.1 Storage of Symbols and Historic Data

Similar to the process in the human brain the perception is used to create the world representation. Perception is a transient process, meaning that it happens in an instant and is updated or overwritten in the next instant. An important question is how the information that enters is preserved and not overwritten by impressions that are more current. Memory as described in section 1.3 requires some form of transformation (coding) in order to provide access to it later on. The process of remembering past events is a complex sequence of reconstruction in the human brain, but is solved differently in the ARS system. In principle, the information originating in perception (that is, microsymbols and snapshot symbols) could be discarded after they have been used to update the representation. Perception has fulfilled its task and provided updates of the real world, which is used to update the representation of the world. Because the flow of symbols between different modules is completely transparent and additionally backed by a database, it is possible to store all symbols for a longer time. It is not necessary to process information and afterwards discard the lower level symbols. Instead, they are written to the database and remain there. By leaving perception information in the system, it is possible to run a long-term analysis (e.g. what has happened today or this week?) and create higher-level modules that use past information for deeper analysis of the data or process perception data. This could include analysis of symbols that could not be fit into place earlier, which can result in the system discarding it, but it can also indicate error conditions (e.g. broken sensors) in the system.

The reason for the difference between a biological system and the ARS system is that the symbols used here undergo only a simple encoding (Figure 8.1) While memories of a human being are consolidated and transferred from short-term to long-term memory in a process that is until now not fully understood (chapter 8), the ARS system deals with pieces of information that can easily be stored and retrieved. However, the comparison between human memory and symbol storage shall not be taken too far here, because both systems operate very differently and the way the human brain stores information is by far more complex than the mechanisms employed here. The ARS system for once stores its perception "as is", meaning just the symbols as they occur, but it also transforms the available information to create and update the representation. A convenient side effect is that also the representation symbols are stored in the same manner, so that the history of representation is available just like the history of perception.

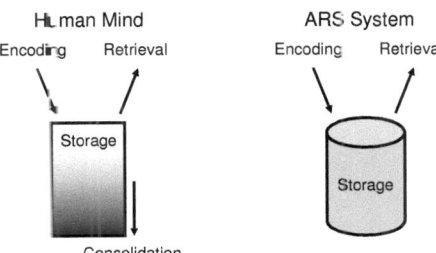

Figure 8.1 Memory of human mind [Sol02] and ARS system

The link between perception and representation also remains intact: when a snapshot symbol has been used to update a representation symbol, this association is noted so that later on the development and evolution of a symbol can be reconstructed. The connection between what the system perceived and how this information was used to build the representation can be important to explain the existence and state of representation symbols. In addition, it can be used for later applications to implement learning and improve the associations between perception and representation. Finally, knowing more about a person or object in the real world (because the information from past perceptions is still available and linked to it) can contribute to better identify the person or object in future.

8.2 Outer World Representation

The representation of the outer world is updated by information originating from sensor information. This process does not happen directly, but goes through several modules that process the data (chapter 10). On symbolic level, this is represented by the creation of microsymbols on the lowest level, followed by snapshot symbols. Representation symbols are created out of snapshot symbols and are the main component of the inner representation.

There is also another source of information for the representation of the outer world, which is found at the other end of the information flow through the system: the system uses the representation of the outer world together with its history to look for anything that requires actions to be taken or represents an important sequence of events. Such a sequence of events is called a *scenario* and is described in section 8.4. Every time a scenario is recognized, the system creates an additional symbol that is also included into the representation of the outer world. If a detected scenario requires the system to take action, the according action series symbols (section 4.1.3) also become part of the representation. This way it is possible to have not only the external events present, but also the reaction of the system. Since the history of representation is stored and not discarded immediately, a later analysis allows seeing how the system reacted to the events in its environment.

8.3 Inner World Representation

The second representation is the *representation of the inner world*. In [Sol02] Solms identifies an inner milieu of the human body that has vital requirements to be fulfilled by the human being. Not being able to meet the requirements of the inner milieu will cause damage to the body and must therefore be avoided. The ARS system has components that are not part of the outside world, but belong to the system itself. Although the system's possibilities to satisfy its own needs (which can be translated to needs for power supply, communication and the like) are very limited, it is still sensitive to define an inner world (and thus a representation of the inner world). The "inner milieu" of the system contains all the components that belong to the system, that is, all sensors and actuators, communication lines like fieldbusses and other networks, and the computers that run the software of the system. Wherever possible the system gathers information about the state of the system components and considers this information for its applications. Therefore, the representation of the inner world is built in analogy to the inner milieu of the human body.

Three things need to be stated: for once, the ARS system is not designed to keep itself alive and use its environment to fulfill its needs. Instead, the system is designed to fulfill needs of human users and keep up operation. At the level of hardware, the system in fact cannot fulfill any of its needs[39] – if a sensor, actuator or communication link fails, it has no means to repair the component. Therefore, the inner world is understood as a collection of system status data used for maintenance purposes. Second, any needs that result from the inner world have to be fulfilled by, for example, a human operator. In other terms, the inner world as it is defined here contains data that is used for two purposes: by human operators for system maintenance and by the system itself to consider any known failures of system components. Finally, the system is not designed to have full information about the state of all its components. It is, for example, likely that a sensor fails without the system receiving information about the failure. Therefore, the system has "white spots" in its knowledge about the state of its components, where it will by default assume correct operation.

8.4 Scenario Evaluation

Using the world representation and its history, the system identifies sequences of events that are of importance; these sequences are called scenarios and are defined for different applications in chapter 11. Scenarios are a mechanism for recognition of events that are not limited to an instant (or short time period), but stretch over longer periods of time. While lower animals are not equipped with higher cognitive functions and live without a concept of time (at least for periods longer than a few seconds), higher animals and especially humans depend on memory of events in the past to evaluate situations. Such behavior is modeled by scenario recognition. Typically, a scenario lasts from seconds up to a few hours; it requires a certain sequence of events to take place while certain conditions have to be met. Since the world representation is the base for scenarios to be recognized, these conditions and sequences are available in representation symbols and their properties.

When a scenario has been recognized, the system can react to this scenario by triggering an action. As opposed to classical control system, scenarios are based on symbolic information, which implies that reactions are not dependent on sensor values, but rather consider the context of the situation on symbolic level. Instead of defining sequences of sensor values that need to be met, a sequence of symbols on representation level provides for abstract definition of scenarios.

Recognized scenarios cause the creation of scenario symbols; this symbol (together with any actions that are triggered by the scenario see section 8.5) becomes part of the representation, thus allowing for a full view of the history of the system, its

---

[39] "Needs" in this context relate to the fact that the system requires repairs or maintenance; the emotions mentioned in section 1.4, could be used, when the system has the "need" to react on a situation; these are however two different concepts and the word "need" is be avoided elsewhere in this work.

environment and the reactions of the system by querying the history of the representation.

8.5 Action Subsystem

The ARS system is mainly observing the real world, its main focus is not the interaction with the real world. Still it has actuators that it uses to influence the real world, which is available to the system by the representation of the outer world. Once a scenario in the representation is found, which requires action, the *action subsystem* is triggered to initiate an action. The concept of symbolization is applied to activity of the system as well. There is only one trigger for action in the system: a scenario that has been recognized and that requires an action to be taken (see section 8.4). While a human being has multiple means to react on influence from the real world, e.g. very fast by reflexes or comparatively slow by cognitive analysis of a situation, the ARS system uses only a single mechanism for acting and reacting. If necessary, this mechanism can be accelerated to behave similar to a reflex by reducing scenarios to single symbols that need to be present. Due to the general passiveness of the system (it observes rather than interacts), it is not sensible to implement multiple mechanisms for interaction with the environment.

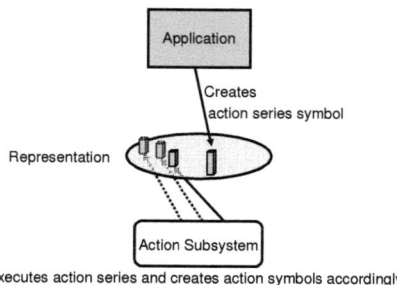

Figure 8.2 Based on an action symbol created by an application, the action subsystem executes actions

When an action needs to be taken (e.g. a room needs to be heated), the action subsystem is instructed to execute a sequence of action. This sequence manifests in the inner representation of the outer world as action series symbols and action symbols[40]. Figure 8.2 shows an action series symbol that is created by an application, because a scenario has been recognized (not shown in the figure). The action subsystem starts processing the actions that are required for the action series symbol; as the actions are executed, action symbols are written into the representation. Thus, the interface to the action subsystem is the action series symbol: such a symbol is sent

---

[40] As stated in section 10.1, the representation consists not only of representation symbols, but also contains scenario symbols, action series symbols and action symbols.

to the subsystem, which then triggers the according set of action symbols. This way the application that set the action series symbol does not have to deal with further execution of the action. The feedback from the action subsystem to the other modules is identical to the communication between all other modules: symbols are written into the representation; in the case of the action subsystem these are action symbols. This way a module can check, if an action is currently going on and modify its behavior accordingly.

9 Reference Applications and Services

The reference applications described in this chapter are taken from different domains, but share a common environment. The main application area of the ARS system is building automation, therefore the system is intended to be installed in a building, either an office or a private home. In this environment the applications of the system are designed to fulfill different tasks, each of them one specific task, but sharing common subtasks, like determining the position of a person. In order to demonstrate the approach of the system, a set of applications were selected from different domains, including person surveillance, geriatric care and child safety. All applications need information about the environment that they operate in.

The applications described here share a common setup of the environment: a number of rooms on one or more floors are the "world" for the applications; this environment has a layout that is identical for all applications – all applications work in the same world[41] simultaneously. The sensors that are used are also identical, and mounted in the same positions. In this way, the symbolization mechanism shares a common set of symbols[42].

Applications run in parallel and their tasks (or subtasks) overlap, therefore the following structure has been introduced: there are *application services* available that can be used by applications; these application services have three basic tasks: person tracking, object tracking and recognizing human activities. An application that needs an object to be tracked passes this task to object tracking. The description of the object and all necessary algorithms need to be programmed into object tracking for the new object, meaning that some parts of object tracking are always different for each object; still, a set of algorithms and methods remains identical and can be used for every different object.

Obviously, the system has only a limited understanding of the world it perceives. The fact that cameras are installed does not automatically imply that the system is able to process all the information in a way that is similar to a human operator observing and

---

[41] Most of the applications operate in the outer world as described in chapter 8, while maintenance applications, which are responsible for the correct operation of the system itself, operate on the inner world.

[42] Although not all symbols have to be present in all applications.

evaluating a camera image. For example, suppose that a dog enters the room. Since the system has no initial concept of a dog, it could possible perceived the dog as a "person" (or, at best, as a "child"). Hence, the system is bound to make incorrect decisions if it is confronted with facts or events that are outside the scope of its capabilities. This system attribute is intentional, since it does not form part of the task that needs to be fulfilled. If we introduce a new application, which makes it necessary to distinguish animals from persons, the knowledge of the system will have to be extended.

Applications operate only on representation level, as shown in Figure 4.1. All the processing being done in applications has impact only on representation layer (including scenario and action series symbols). However, the design of an application is not limited to representation level, because the symbols available there depend on symbols of both snapshot and microsymbol level. Therefore implementing a new application most likely requires changes in all modules, at least by defining new symbols on each level.

9.1 Environment Model

Aside of sensor information that builds the base for perception and thus understanding of the current status of the world, the system needs information that cannot easily be obtained by evaluating sensor information. There needs to be information about the environment available in the system, which allows it to know about the layout of the building, the location of doors, the location of objects that cannot be perceived and so on. This knowledge is used when deciding the whereabouts of persons (e.g., in which room a person is). The information does not originate from sensor information (rather in terms of blueprints and building layout) and is described in detail in section 6.1. It is important to understand that the model of the environment strongly depends on the application of the system: since the reference applications described in this work are all located in the area of building automation or private homes, the environment model reflects this fact by representing office buildings with a determined structure that is covered in section 6.1. The environment model includes all static equipment that serves no purpose for the perception of the system other than defining the layout of the building. This includes all inventories that do not have sensors or actuators, which cannot be controlled by the system. The stove, for example, is an important "active" object for the application Child Safety (section 9.6) and is therefore not part of Environment Model, but rather of Object Tracking (section 9.2).

Because the ARS system does not only deal with perception of objects and persons, there is also a symbolic representation of rooms in the system, for one specific reason: the application Comfort (section 9.8) is responsible for controlling climate and lighting in the building. To do so, it operates in a similar fashion as the other applications, by processing symbolic information and creating and modifying symbols. Therefore, the Environment Model represents the room layout in a symbolic fashion, with additional information about climate and lighting.

9.2 Application Service: Object Tracking

This application service supervises the whereabouts of objects. In a given room or building the system monitors a set of objects; updates and changes like an object being moved or taken by a person[43] are tracked by the system and stored into the history of the according representation symbol. Object Tracking is therefore able to tell the location of an object and – if applicable – the owner of an object.

The Object Tracking service[44] is based on the ability of the system to perceive objects, which is done in the perception module (Object Recognition, see section 7.2). The service builds upon snapshot symbols that have been created by Object Recognition. Since Object Tracking operates on the representation of the outer world, there are not many symbols present and new symbols are only rarely created. Instead, the existing symbols have to be matched with incoming symbols from perception and transformed into updates of the perception symbols: Perception will continuously report that, for example, a chair has been perceived at a certain position. The task of Object Tracking is then to find a chair on representation level and make sure that it is still the same chair as the one, which is reported by the current perception.

Object Tracking is an application service, because there are different applications that need to know about the whereabouts of certain object. While each application is only interested in a certain subset of objects and no application will likely use all objects that this application service knows about, it is reasonable to group the task of object recognition into one application service. For once, the relevant object for two applications might overlap (e.g., two applications are interested in the whereabouts of chairs). In this case, the same recognition process would have to be done twice, and would thus be redundant. Second, the mechanisms for tracking are similar for different objects and therefore have to be available only one, if the tracking of objects is grouped into one application service.

Not all objects need to be perceived visually; the stove, for example, which is needed in the Child Safety application (section 9.6) is a fixed part of the inventory, still it can change its status (being hot and thus potentially dangerous or not) and is therefore also included into Object Tracking.

9.3 Application Service: Person Tracking

By making use of light barriers and motion detectors, pressure sensors in the floor, door contacts, cameras and other sensors, the system is enabled to know where people

---

[43] Or the lack of changes, meaning a permanently present, unchanged object.

[44] Note that applications and application services are written with capital letters in order to separate them from common terms, e.g. when referring to object tracking in general.

are in a building. People are considered anonymous, which means that the system has no additional knowledge about their identity in terms of name, social security number, or similar identities[45]. The system is able to provide information about a person's current and past location, so that the path of a person through a building can be tracked and monitored.

9.4 Application Service: Human Activities

Aside of knowing where persons are, an application needs to recognize what persons do. The application service Human Activities builds on the application service Person Tracking, Object Tracking and Environment Model and analyzes the basic activities of persons, such as walking, running, sitting, or talking. Human activities are part of the representation of the outer world and rely on a set microsymbols and snapshot symbols. Figure 9.1 gives an overview of the relations between the basic activities and their dependence on other Application Services. The arrows indicate contributions from other application services to the creation of activities. Human Activities is responsible for the basic activities talk, lie, sit, stand, walk and run. Activities are not exclusively managed by the Human Activities application service. Additional activities can be added by other applications (Figure 9.2). In the system described here, the application Person Surveillance (section 9.5) adds two higher-level activities work and meeting shown in Figure 9.2.

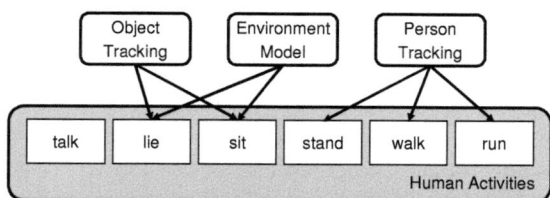

Figure 9.1 Relations between basic activities in Human Activities and other application services

Activities are recognized based on different factors. For the activities stand, walk and run the system first needs to identify that a person is within its range of perception. This is the task of Person Tracking (see section 9.1). Based on the current position and the history of positions the system is able to decide, which of the three activities fits best in the current situation. The creation of the activities stand, walk and run does not need any additional microsymbols, since it relies on the position information provided by Person Tracking. On the level of snapshots all three activities are present as properties in the symbol osnPerson. They are created from the difference between the last and the current position of a person and information that

---

[45] See also the comments about privacy in section 9.5.

provides the direction of movement (e.g. by triggering adjacent floor sensors in a defined time span). Two thresholds in the speed of movement decide which activity is most accurate. The lower threshold between sit and walk is close to zero, the threshold between walk and run is selected so that perception can accurately select between persons who walk or run. In the representation of the outer world, the three activities are also present in the symbol person for representation (crpPerson), but the creation of such an activity does not only respect the last position, but rather the history of positions. Since perception might be incorrectly assigning activities or change between two activities at a high rate, the representation uses more data to determine the activity more precisely.

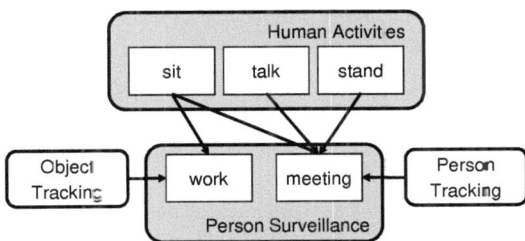

Figure 9.2 Higher-level activities "work" and "meeting" depending on basic activities

The activities sit and lie need both Person Tracking and either Object Tracking (section 9.2) or Environment Model (section 9.1). The two factors that are needed are: a person that does not move and an object nearby, which is feasible for a person to sit or lie down. Since Object Tracking only knows chairs in the work described here, it can only contribute by providing the position of chairs. The Environment Model can also contribute feasible seating object, if they are declared as such. Similarly, a person can lie only on an object that is known to be a bed (either by Object Tracking or Environment Model). With the activity lie the system also has to consider the option of a person lying on the floor, which is a possibly dangerous scenario of the Geriatric Care application (see section 9.7) and is treated specially in the according application. When the two conditions are met (still person, feasible object for sitting or lying nearby), the according activity is created. It is present in the snapshot symbol for a person (osnPerson) as well as in the representation symbol for a person, only that the representation symbol filters for plausible activities using not only current perception, but also historical data.

The activity talk is based only on aural information originating from the microphones in the system. It does not depend on contribution of other application services, since all stream data bypass the way of regular sensors (which are written into to the sensor database, see section 12.4.1). Processing of stream data is done directly in the sensor or an attached networked device, should the sensor not be capable of doing the necessary processing. The activity `talk` relies on the ability of the system to determine human sounds and discriminate those sounds from environmental sounds (e.g. wind) or sound created by machinery (e.g. air condition). The perception of this activity uses the microsymbols from aural sources and creates the talk activity in the person symbol on snapshot and representation level (`osnPerson` or `orpPerson`, respectively).

There are two higher-level activities that are derived from the basic activities (Figure 9.2): in an office environment as assumed in this context, the activity `work` is important to be recognized by the system. The abstracted model of a person doing desktop work is defined as a person that sits and a table being in front of the person. Therefore, the activity work relies on the activity `sit` and additionally needs Object Tracking or Environment Model to detect a table in correct spatial relation. Again, the activity `work` is present on both snapshot and representation level.

In the same context of office buildings, the activity `meeting` is defined as two persons having a talk or discussing something. Aside of the activity `talk` we therefore need Person Tracking to provide us with the information of another person being present in the same room. The spatial correlation of two people is sufficiently described by either "being in the same room" or "close to each other". Additionally, both persons have to either `sit` or `stand`.

Aside of the perception of activities the application has the task to classify people. There are three different classes defined: a child, a grown-up and a senior. This classification is needed for the two applications Child Safety (section 9.6) and Geriatric Care (section 9.7). The decision whether a person is a child or a grown-up is done by the height of the person (see also section 9.6) and can be improved in future implementations. Seniors can per se not be detected by the system; the system instead has to know a priori about persons, who need special treatment; this information is also available to Human Activities, thus it is responsible for classifying a person as one of the three defined classes. Since Human Activities is expected to have more detailed analysis of persons and their activities than any other module in the system, the task of person classification is assigned to it, as it will emerge as a side result during analysis. The information resulting from classification is stored in a property called `lifeStage` in the representation symbol `orpPerson`.

9.5 Application: Person Surveillance

Person Surveillance deals with the whereabouts of persons and their activities. The term surveillance may be misleading in the sense that the right for personal privacy might be compromised by the system, but has been chosen nevertheless for the following reason: systems that are available today do in fact have the potential to be abused for observation. Closed Circuit Television (CCTV) systems with the possibility to record the events that are monitored contain all the available information from a camera image. The Person Surveillance system that is intended here goes the opposite direction: since the system itself is able to process the available information, the sensor information – and especially the camera images – does not have to be stored in order to reconstruct the events that have happened. Instead the system processes all sensor information, symbolizes them and creates a world representation that only contains information that is comprehensible for the system, for example, where persons are or which activities they posses. These activities are limited to the applications that the system implements. The information that two persons have met (i.e. had a meeting) may be available, while information about the topic or what has actually been spoken, is completely stripped from the symbol-based world representation.

This application is strongly based on the combination of two services: Person Tracking and Human Activities. It combines them to gain a view of the world where persons are shown together with their activities. The Application Person Surveillance does not create scenario symbols, since the information that is relevant for this application is already represented in the symbols of the inner representation of the outer world. It relies on position information provided by Person Tracking (section 9.3) and the basic activities provided by Human Activities (section 9.4). Using this information (and other information available from Object Tracking and Environment Model) the application Person Surveillance creates higher-level activities, derived from the basic activities. The two activities that are in the scope of this work are `work` and `meeting` (see also section 9.4).

Another task of Person Surveillance is to know, whether a person possesses a classified object. To do so it cooperates with the application service Object Tracking to know about the location of objects. When objects disappear or remain near a moving person, it is assumed that the person carries the object in question. By setting the property `possessor` of the object, this information is stored and available for later reference.

9.6 Application: Child Safety

The second application considered is a child safety system. The system recognizes when a particular person is actually a child, and can monitor and guard the actions of the child. When it appears that the safety of the child may be compromised due to a hazardous situation, the system alerts a (human) supervisor. This is done by creating a scenario symbol indicating that a dangerous scenario has been detected.

The decision that a particular person is actually a child is based on diverse sensor information, similar to other mechanisms in the system. On microsymbol level, this includes the use of camera images that allow a decision regarding the height and shape of a person, as well as information from light barriers that are mounted at different heights. All this information is collected and processed to set the property `lifeStage` in the symbol person, where children are identified (and separated from grown-ups or seniors, respectively). Situations that are classified as hazardous are: a hot stove, or a child climbing on a table. There are additional conditions and criteria that have to be taken into consideration. For example, a situation is only classified as hazardous if a child is alone and unattended (meaning that no adult is nearby) and the stove is indeed hot.

9.7 Application: Geriatric Care

The third application is concerned with a geriatric system to care for elderly people. In this case, the system recognizes when an elderly person collapses or faints. Furthermore, the system makes use of Object Tracking by providing information of the whereabouts of specified things. This is intended as a support for absent-minded seniors, who require assistance in finding necessary items such as glasses or books. However, the main application of Geriatric Care is still the safety of elderly persons. Aside of detecting collapses, the system shall also know the position of an elderly person and communicate this position to personnel. This way seniors, who are lost and are not in their usual environment, can be found by support of the system. The necessary position information comes directly from Person Tracking. Additionally, the application identifies a senior, who is lying or sitting in a room, where the temperature has dropped to a possibly dangerous level (e.g. because the senior opened the window and forgot to close it before going to sleep).

9.8 Application: Comfort

Controlling room climate and lighting in an office building or private home is a classical task of building automation. The ARS system with its extended understanding of persons and their activities can support HVAC (heating, ventilation, air conditioning) applications as well as illumination. The difference to the other applications described in this chapter is the permanent use of actuators: while other applications rely on observing objects and persons without direct interference, the Comfort application permanently controls lights and air conditioning systems. It is not the goal of this reference application to implement a new set of algorithms for efficient comfort control, but to show how the use of sensors of different industries can operate together to achieve a better result. This is still not common in today's building automation applications ([Lon02]), where each industry uses its own set of sensors and actuators to fulfill its tasks. When the number of sensors increases this will result in more and more redundant sensors, where redundancy is not used, but ignored by other applications. Since the system described here is designed to use many different sensors to extract different kinds of information for different applications, it fills the gap of today's lack of inter-industry interoperability.

The tasks that have been selected for the Comfort application include setting the room temperature to a comfortable level[46] depending on the presence of persons; the lighting level is set accordingly in order to implement not only comfort for persons working or living in the environment, but to also use energy efficiently when rooms are not used.

9.9 Outline of Reference Applications

Figure 9.3 sums up the previously described applications and application services. The services provided by the three application services (the environment model is not explicitly mentioned here, since it is an integral source of information for all application services) are used by the applications to fulfill the according tasks. The structure avoids redundancies and thus inconsistencies, because different applications need similar information. By the set of application services that is defined here, each task has to be done only once and not separately for each application

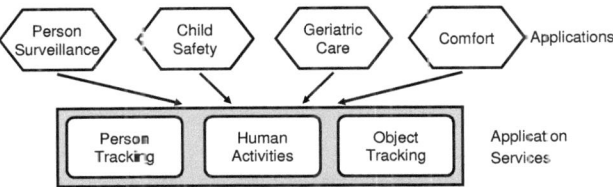

Figure 9.3 Applications depending on Application Services

A more detailed view of the interconnections between applications and application services is given in Table 9.1. The information provided by Person Tracking is needed by all applications, since it is not possible to operate on a person symbol and its properties without knowing the position of the person. The different basic activities provided by Human Activities are also needed by all applications, where Person Surveillance uses them to create higher-level activities. Objects that are in the environment are used for different purposes in the applications, either for higher-level activities (Person Surveillance), to detect possibly dangerous situations (Child Safety) or to support elderly persons (Geriatric Care). The application comfort has one special communication path, since it uses not only the basic activities, but also the higher-level activities created by Person Surveillance; therefore it depends not only on application services, but also on the output of another application. This behavior is expected, especially when seeing it in the light of future extensions. Since applications are thought of as functions or duties that the system has to fulfill (or, when compared to the human mind: a goal that a person pursues), it is likely that the results of one of the tasks contributes to the fulfillment of another task and that information, which is available to the system, is used in many different, cross-linked

---

[46] This is done by employing already existing HVAC systems and not by implementing a new system.

paths. The implementation of this cross-link is solved by sharing a common data pool, that is, symbols, which are consistently available throughout all applications and application services.

	Person Tracking	Human Activities	Object Tracking
Person Surveillance	yes	stand, sit, walk, run, talk	table, chair, placeholder object (orange ball)
Child Safety	yes	stand, sit, walk, run, lie, classification of children	all dangerous objects (table, stove)
Geriatric Care	yes	stand, sit, walk, run, lie, classification of seniors	glasses, keys (represented by placeholder object)
Comfort	yes	stand, sit, walk, run; from Person Surveillance: work, meeting	

Table 9.1 Requirements of basic functionalities by each application

This pool of symbols is stored in a database and accessible by a communication infrastructure (see section 12.3). In principle, every module can access all information, since there are not access restrictions implemented on communication level. Still, applications access different sets of symbols than applications services as shown in Figure 9.4. While application services are responsible for creating and updating lower level symbols (i.e., perception and representation symbols), the applications use these symbols to create scenario symbols (which, consequently, cause the creation of action and action series symbols)[47].

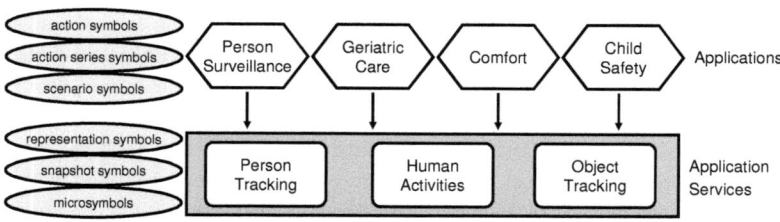

Figure 9.4 Applications and application services access different levels of symbols

It is also important to understand that every application needs the whole system, which means that the introduction of a new application (i.e. giving the system a new goal that it should pursue) might require changes on all symbol levels. This is because a new application, e.g. adding gesture control to the comfort application or voice

---

[47] Note that the top three categories are also considered being part of the representation; they augment the representation symbols and become part of the representation (section 4.1.3).

control for the whole system, has new requirement to perception of the environment. To understand gestures the system needs a much more fine-grained visual image of the person that interacts with the system, therefore more microsymbols, snapshot symbols and properties for representation symbols are necessary. Figure 9.4 does not show these relations, it rather shows the hierarchy between applications and application service and their allocation in the different levels of symbols.

PART III: Concepts and Results

This part describes the concepts and methods, which lead to a prototype of the model that has been designed. Sensor data is obtained in two different ways: a real-world implementation is used to get accurate sensor data, while a simulated environment is used to generate big amounts of data, which would be infeasible to generate in a real-world installation. The interface to these two different data sources is identical and is implemented as a database for sensor values. The symbolic alphabet that builds the core of the model is described and assigned to the reference applications. The communication infrastructure, which is necessary to run the distributed system is explained. Finally, visualization and simulation are described.

10 Module Description and Data-Flow

This section describes the system design in terms of information flow and software design. The description consists of two parts; the first part contains the description for the data flow that leads to the inner representation of the outer world. It explains all modules necessary to get there and showing how data are exchanged between the modules.

The second part describes how the representation of the inner world is created. The modules and data flow paths are very similar to the ones in the first part – an intentional design decision. The two representations of the inner and outer world shall be created in similar ways (the system shall use the same mechanisms no matter whether it "looks" inside itself or into the real world). In fact, the modules are partly identical, using only different sets of symbols and associations to combine them. Still it is important to separate the two representations with all their data paths and data records.

10.1 Modules for Outer World Representation

The source of real world information is the sensor data, which is collected in a database scheme described in section 12.4.1. The first processing step is done in a module called *microsymbol factory* (Figure 10.1); it creates microsymbols based on sensor data and writes these microsymbols into the database scheme described in section 12.4.1. Microsymbols are the information source for creating *snapshots*. Snapshots contain the current state of the world; they are created in the *perception module* and use the information from microsymbols to create a snapshot that is eligible to become the next update of the world representation. The step from snapshot symbols to representation symbols is done in the *representation module*. The task of the representation module is to take the snapshot symbols that have been created and evaluate them. If they fit as a plausible continuation of what has been the world representation so far, the representation is updated. It may discard snapshot information that appears not plausible and it may insert information that was not perceived by the perception module.

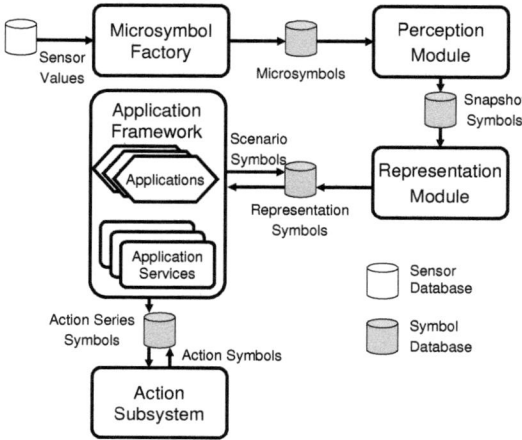

Figure 10.1 System components and data-flow for outer world

The representation with its different symbols is the highest level of symbolization that is created in the ARS system. To get such a representation the information from sensors has to be processed by all modules (with additional symbols added by the applications whenever a scenario is recognized, see section 8.4 and 8.5). As opposed to pure perceptions (which manifest as snapshots) the representation contains information that is augmented by knowledge about "how the world works" (section 1.2.1) and can be used in combination with additional knowledge to which the system has access. Thus it represents information of higher quality – a lot of condensed information is used and combined with knowledge.

Based on the representation of the outer world and its history the system looks for scenarios (see section 8.4). A scenario represents a (temporal) sequence of events that were perceived by the system. An application (section 9) that runs in the *application framework* therefore has to define one or more of these scenarios[48].

Only when an application recognizes a scenario an action is triggered. A common action is to notify a human operator that a certain scenario has taken place, other actions include control of actuators by using the *action subsystem* (section 8.5), such

---

[48] An example taken from the geriatric care application: a senior goes to sleep, leaves the window open and the temperature in the room drops to a possibly dangerous level.

as adjusting the room temperature or lights. There is no "shortcut", which would allow faster reaction by, for example, triggering directly on a single sensor input[49].

Modules are designed to work as stand-alone pieces of software, therefore each interface they have to another module has to be defined according to the specifications in section 12.3) and use only the mechanisms described there. This way the system can be designed in a modular way and extended without knowledge of all system components. This is vital, since for once it is expected to continue development on the system for future projects; second, the different components might consume considerable resources (at least in the prototype implementation), which could lead to bottlenecks, if the complete system would have to run on the same machine. This modular design rule is also a reason why the "reflexes" mentioned above are not implemented. If, for extended application scenarios, they appear sensitive to be implemented, a new interface between application and sensor database would have to be defined.

The exchange of information between the different modules as shown in Figure 10.1 is backed by databases: the sensor database (section 12.4.1) and the symbol database (section 12.4.2). The microsymbol factory creates microsymbols from sensor values and writes them into the symbol database. From there the perception module reads microsymbols and creates snapshots that are written into the symbol database. Similarly, the representation module reads new snapshot symbols and updates the representation symbols in the symbol database.

The representation symbols that make up the representation of the outer world are examined by the applications running in the application framework to find scenarios. If a scenario is found, the application writes a scenario symbol back into the database. This implies that the representation module has to query the database, if it wants the complete representation including the scenario symbols.

When an application has recognized a scenario that requires actions to be taken, it does so by writing an action series symbol into the symbol database; this is read by the action subsystem, which executes the action series. The action symbols that are required by the action series are created by the action subsystem and written into the symbol database.

10.2 Modules for Inner World Representation

Figure 10.2 shows the modules and data flow for the process that creates the representation of the inner world. The modules are named identical to the definitions in section 10.1 and have identical functionality. The sensor database on the left is now

---

[49] This would be an analogy to a reflex in a human being – a fast system reaction that circumvents the higher cognitive functions, similar to the subsumption architecture described in section 1.1. Although possible, it is not in the concept of this system in order to avoid unnecessary complicatations. The one existing mechanism that triggers actions should suffice the requirements of the applications.

the source for data coming from the components of the system. This includes status information of sensors, reports about failures in the network, error messages from the communication system, actuator failures and so on. The rest of the process up to the representation module is identical. An instance of the microsymbol factory applies its methods and algorithms to create microsymbols as pieces of perception of the inner world; the perception module assembles these microsymbols to create snapshots of the inner world and the representation module builds the representation of the inner world.

There is no dedicated action system that uses the representation of the inner world. If the representation module finds failures in the system, it writes an according symbol into the symbol database. The substitute of the action system (not shown in Figure 10.2) is a periodical scan for error symbols; if an error is found, it is reported to a human user.

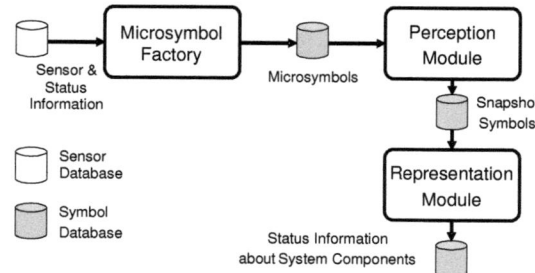

Figure 10.2 Data-flow for representation of the inner world

11 Symbol Definitions

Symbolic processing requires a set of symbols; this alphabet of symbols is presented here, assigned to the reference applications presented in chapter 9. The symbols belong to different applications and are located on different levels of symbolization, they are listed with their properties, of which some may be modified by different applications, thus a symbol does not exclusively belong to one application.

In this chapter the symbols defined, which are then used during operation. While the definition of symbols is static and remains unchanged during the lifetime of the system, they are nevertheless repeatedly instantiated during runtime of the system. Consequently, symbol instances are destroyed at the end of their lifetime. To avoid the lengthy term "symbol instance" in the following text, the differentiation between symbols and symbol instances is not mentioned; whenever symbols are created or destroyed, this refers to instances of symbols, not to the definition of symbols.

11.1 Symbol Categories and Naming Conventions

The different classes or levels of symbols that have been introduced in section 4.1 are used in the implementation to create the symbolic information-processing scheme that is a vital concept of this system. The symbol hierarchy starts at microsymbols, which are created only out of sensor data, followed by snapshot symbols, which contain a momentary view of the world as the system sees it. Snapshot symbols are created by evaluating microsymbols and build the base for representation symbols. These three layers build the "passive" part of the system, where the system observes its environment (or inner world) and creates a representation of this world. The active parts of the system include creating scenario symbols, which means that the system was able to identify a scenario (which usually requires some action to be taken). Action series symbols are made up of action symbols and are – together with the scenario symbols – also part of the world representation. They are created by the system whenever interaction with the world is required. this includes notification of a human operator as well as modifying room climate settings.

Symbols can be created and destroyed; lower level symbols typically have a shorter lifespan than high-level symbols. A symbol can have one or more properties, which contain further information relevant for the symbol. Properties can be updated individually by different applications. However, if a symbol is destroyed, all according properties are destroyed as well, since properties cannot exist alone.

Symbols are commonly specific to an application or application service. This means that implementing a new application requires new symbols to be created on all levels[50]. However, symbols do not need to be unique for each application; especially on representation level the number of symbols is low, meaning that different applications can work on the same symbol, e.g. by modifying different properties.

In order to avoid ambiguities between symbols of different levels[51] some naming conventions were adopted, which indicate what kind of symbol is used. Aside of different levels the system also knows the inner representation of the outer world and the representation of the inner world (see section 8.2 and 8.3). Although the symbols existing in both representations are not similar, the affiliation of a symbol to inner or outer world representation is also indicated.

Symbol level and inner or outer world representation are indicated by a prefix to the symbol name: first, the representation is pointed out by either an "o" for outer or "i" for inner world representation. The following two letters provide the symbol level:

- ms for microsymbol

---

[50] Implementing new applications can be compared to the system "learning" new skills.

[51] A person symbol created on snapshot level is different from a person symbol created on representation level: snapshots only exist for an instance, while representation symbol persons only have their properties updated.

- `sn` for snapshot symbol
- `rp` for representation symbol
- `sc` for scenario symbol
- `ac` for action symbol
- `as` for action series symbol

The microsymbol for a footprint originating from a tactile floor sensor is therefore called `omsFootprint`, the representation symbol for a person is `orpPerson`.

If ambiguity is still possible, e.g. because different sensors create a similar microsymbol, but with a slightly different meaning, the type of sensor creating a microsymbol can be appended to the end of the microsymbol name:

- `lb` for light barrier
- `md` for motion detector
- `ts` for tactile floor sensors
- `irH` for infrared height sensor
- `mi` for microphone
- `ca` for camera

For descriptions that are more detailed this suffix may be extended, for example, `lb1` for light barriers of type 1, `lb2` for light barriers of type 2, and so on.

11.2 Environment Model

Aside of perception, which is responsible for persons and objects, there is additional knowledge in the system. The environment, meaning the layout of the rooms in the building, is not perceived by sensor information. The ARS system is not designed to understand and reconstruct the layout of rooms; this information is available a priori (see section 6.1). Still, some of the information about the environment is available symbolically, by means of representation symbols. It is used for the application Comfort (section 11.9), which controls climate and lighting in the building. The Environment Model defines only one representation symbol.

`orpRoom`: A static representation symbol that is created upon start-up of the system and never destroyed. It represents one room in a building (where "room" means all places where persons can be, including halls).

Properties

`orpRoom` has two properties, `temperature` and `humidity`, which contain information about the climate in the room and `lighting`, which represents the illumination level in the room.

11.3 Application Service: Object Tracking

The task of the application service Object Tracking is to identify and follow objects that are not static, but can be moved around in the environment. One special object is the stove, since it is actually part of the layout and cannot be perceived by the system, but it is still part of Object Recognition, since it can change its status from harmless to possibly dangerous.

The objects of interest defined here are a table, a chair and an orange ball. Table and chair are used for the child safety application and for general purposes of person surveillance (i.e. position determination and activities like "working", which requires the presence of a table). An example for perception of a chair is given in section 7.2.1, perception of tables happen accordingly. The orange ball is a substitute for more complex objects, which require deeper integration of image recognition methods. However, covering image recognition as a separate research topic is not the goal of this work. The orange ball is an object that is commonly used in, for example, robot soccer, where autonomous robots determine the location of the ball and score a goal. It is easy to detect, since it has a distinctive color and a shape that is identical from all side. Therefore, it is used here as a placeholder for other, more complex objects, if the focus is on the application around an object and not the perception of the object itself. Such an application is, for example, the support of elderly people in finding lost objects (glasses, keys, books and the like, see section 9.7). Although it is a complex task to detect, for example, glasses in a real-world environment, the application itself is modularized and can be interfaced and separated from the object recognition itself. More advanced image recognition methods can later substitute the orange ball object by more sophisticated objects with little interference to the application.

Object recognition relies heavily on the sensors that are suited best for the task: cameras. Since it is hard to perceive different objects by use of sensors like motion detectors or tactile sensors, the focus of object recognition is on visual perception. Cameras provide a constant stream of images and are therefore treated specially in terms of data storage: compared to other sensors the amount of data originating from cameras is considerably higher and cannot easily be stored in a database as described in section 12.4.2. Therefore processing of video data is done differently in the lower layers of perception: the input stream is processed in a separate application (that may run on a separate machine). The whole process of object recognition is contained in a piece of software, which has the implication that recognized objects are transformed into symbols directly (see also the comments on privacy in section 9.5) Aside of this peculiarity the flow of symbols is identical: Object Recognition creates symbols and writes them into the database, where they can be used by other applications or application services, especially the Object Tracking application service described here. The symbols created by Object Recognition are defined below.

As stated earlier, these objects represent a certain set of generic tables or chairs without the requirement of being able to perceive every possible object that belongs to these classes. In the real world, there are many different types of chairs, which the system is unable to perceive as such; the same applies for tables[52]. By adding different descriptions of the same class of object (i.e. more different tables or chairs), the system becomes enabled to perceive a more complex environment. A prototype chair is described in section 7.2.1; the definition of a table describes a regular office table, with a rectangular desktop and four legs, one at each corner. Accordingly, the property *dimension* consists of a length and a width and the only other property needed (aside of the position, which is a common property of all objects) is the *height*, at which the desktop is located. The chair needs some additional information: aside of the height of the chair seat the *orientation* is also important. This is done by providing the position of the chair back.

11.3.1 Snapshot Symbols

osnTable: A table object on snapshot level, which contains the properties position, indicating the center of the table, dimension, which contains the length and width of the table and height, which contains the height of the desktop.

osnChair: Similar to the table, this symbol represents a chair with its properties position, containing the center of the chair, seatHeight for the height of the chair seat and chairBack, containing the position of the chair back.

osnOrangeBall: The orange ball, which is used as the placeholder for other objects has just t property, position, containing the position of the ball.

11.3.2 Representation Symbols

The snapshot symbols created by Object Tracking can be mapped directly onto the representation symbols of Object Recognition; they build the base to update the whereabouts of representation level symbols.

orpTable: A table with its properties position, indicating the center of the table, dimension, which contains the length and width of the table and the height of the desktop, contained in the property height.

orpChair: Similar to the table, this symbol represents a chair with its properties position, containing the center of the chair, seatHeight for the height of the chair seat and chairBack, containing the position of the chair back.

---

[52] The orange ball object is different, since it is a placeholder that is defined as being unique in its shape and color. Chairs and tables do not have a possessor in this model.

orpStove: The stove is only present on representation level, because it does not have to be perceived visually. Its existence and position is given to the system as a priori knowledge. The task ob Object Tracking is to track the changes of its status: when the stove is not turned on and cold, it is harmless, which has influence on the Child Safety application, described in section 11.7. A hot stove is a possibly dangerous object and the information about this possible danger is contained in the property isHot.

orpOrangeBall: The orange ball object has a new property on representation level: the possessor. This is a reference to a person of which the system assumes that it currently possesses the object (e.g., a person has taken a book). The application Person Surveillance (section 9.5) sets this property by observing symbols on snapshot and representation level. If an object remains unchanged (e.g. a book lying on the table), the system knows about this fact. If a person comes close to the book and the book can afterwards not be located again, the system assumes that the person now possesses the book. Even if the system is not able to detect all the details of the scene (i.e. the person actually taking the book), the assumption is still valid (with a certain credibility). This shows one of the core properties of perception: information that is not available by sensory input is replaced by knowledge about "how the world works". Even if the scene described above were perfectly visible from all sides by use of many cameras, the effort to detect the actual act of a person taking an object requires a lot of computational and algorithmic effort, whereas the much simpler assumption of "there was an object, then the person came and the object was gone" benefits from a priori knowledge. The requirements to image processing decrease, because it is not necessary to detect body parts and their spatial and temporal relations; instead, a "differential image" of the scene before and after suffices.

11.4 Application Service: Person Tracking

11.4.1 Sensors

The application service Person Tracking uses the most diverse set of sensors to determine the position of persons in a building. Light barriers, motion detectors, tactile floor sensors, microphones and cameras are used and combined to determine there whereabouts of persons.

- Light barrier type 1: single barrier mounted across floor or hall
- Light barrier type 2: two barriers mounted closely together to detect direction of movement
- Light barrier type 3: set of light barriers mounted vertically above each other

- Motion detector: mounted on a wall at half the distance between floor and ceiling with sensitive area of a wedge; alternatively it can be mounted on the ceiling with a sensitive area of a cylinder
- Tactile floor sensor type 1: single sensor mounted on the floor with a sensitive area of a rectangle breadth time width
- Tactile floor sensor type 2: array of sensors mounted on the floor, each without physical dimension (single point)
- height measuring sensors: infrared sensors mounted above persons (e.g. in a door) to measure the height of a person
- Microphone type 1: mounted anywhere in the building; detect noise in a room
- Microphone type 2: array of microphone able to detect a sound source
- Camera: color camera defined by resolution, frame rate and visual area

11.4.2 Microsymbols

Due to the many different sensors that contribute to Person Tracking there is also a number of microsymbols defined. As is common with microsymbol, there are not many properties in a single microsymbol; instead, the number of microsymbols is important. Symbols with lots of properties are common on representation level, but not on microsymbol level. If a microsymbol has properties, they are listed in the according section.

`omsInterrupted`: This microsymbol requires a single light barrier to be triggered from 1 to 0 (where 1 means "not interrupted"). The timestamp for the creation of the symbols is set, the end timestamp is set to infinity. Once the light barrier is not interrupted again, the end timestamp is set and the microsymbol is terminated (see Figure 11.1).

Properties

`position`: The position for an interrupted light barrier is represented as a line that goes from the start to the end point of the barrier.

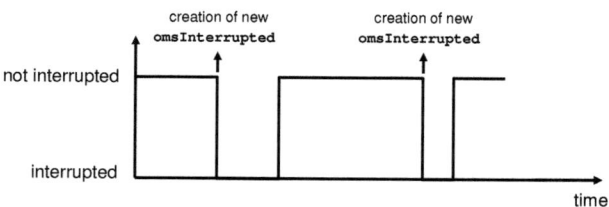

Figure 11.1 Creation of a microsymbol based on de-bounced sensor data

This symbol is also created by a set of light barriers (light barrier type 3), but only once and not for each light barrier of the set. The time representation of this symbol is "period".

omsObstructedLb: If a single light barrier or set of light barriers is interrupted for a time longer than a definable t_{int}, this microsymbol is created with the beginning timestamp being the first interruption. When the light barrier is free again, the ending timestamp is set and the symbol is terminated. While this symbol exists for one specific light barrier, there cannot be a symbol omsInterrupted or omsObjectPassedByLb2 for the same light barrier. Therefore, all existing symbols for this specific light barrier or set of light barriers have to be terminated before. A light barrier that is blocked by an object or a person is not necessarily useless to the system. It may easily happen that a person stops and blocks a light barrier, in this case the light barrier contributes to location information of that person by the microsymbol omsObstructedLb. The time representation of this symbol is "period".

Properties

position: The position for an interrupted light barrier is represented as a line that goes from the start to the end point of the barrier.

omsObjectPassesByLb2: When two light barriers mounted closely together are triggered consecutively, the direction of the movement is derived and written into the property direction. The symbol is created when the first sensor is triggered (meaning interrupted for a time not longer than t_{int}), followed by the second sensor being triggered (which shall not take longer than time t_{gap}) and both sensors resetting to uninterrupted state. Similar to single light barriers this sensor might be obstructed. If either of the two barriers is obstructed for a time period longer than t_{int} the symbol omsObjectPassesByLb2 is destroyed and a symbol omsObstructedLb is created. The time representation of this symbol is "period".

Figure 11.2 Creation of microsymbol omsObjectPassesBy_lb2 using two adjacent light barriers

Properties

`position`: The position for an interrupted light barrier is represented as a line that goes from the start to the end point of the barrier.

`direction`: the direction can have two values that represent left or right. The convention for these values is to look from the side where the receptive device of the sensor is mounted (which might be the device that includes the photodiode and the light source or just the photodiode). Looking from this point into the area where the light beam goes to defines left and right. Naturally, it is not possible to define a global direction for a single sensor; therefore the direction relative to the sensor is used. The above definition assumes that both light barriers have the receptive device mounted next to each other. If this is not the case, an arbitrary definition has to be used.

`omsPersonHeightLb3`: A set of vertically mounted light barriers at different heights is used to determine the height of a person. The symbol is created when all light barriers starting from the lowest position up to a certain height are interrupted. Note that a person walking through such a set of light barriers creates two symbols: `omsPersonHeightLb3` and `omsInterrupted`. If one of the light barriers stays interrupted for longer than time t_{int}, no new symbol `omsPersonHeightLb3` is created, but rather `omsObstructedLb` (only one symbol for all light barriers of the set).

Other patterns than the one for determining the height are ignored and do not create a symbol, except if one of the light barriers stays interrupted for longer than time t_{int}, then `omsObstructedLb` is created. The time representation of this symbol is "instant".

Properties

`position`: The position for an interrupted set of light barriers is represented as a line that goes from the start to the end point of the (parallel) barriers.

`height`: the detected height, derived as the mean value between the two light barriers involved, that is, either the last light barrier that was not triggered or, if all light barriers have been triggered, the predefined maximum size (defined by the layout of the area where the sensor set is mounted).

`omsPersonHeightIrH`: This microsymbol is created by the height-measuring sensor. When a person passes by underneath such a sensor and the height can be reliably determined, `omsPersonHeightIrH` is created, indicating that additional information (the height) about the person is available. The time representation of this symbol is "instant".

Properties

`height`: the height of the person as detected by the sensor like the property `height` in `omsPersonHeightLb3`, only more precise.

`omsObjectMovesMd`: This microsymbol indicates that an object or a person has moved in the sensitive area of this motion detector, which is modeled as a wedge or a cylinder, depending on where the sensor is mounted. Unlike light barriers, a motion detector is not blocked when a person remains in its sensitive area. In order to avoid the creation of too many symbols in a short time, this symbol has a timestamp for begin and end (like `omsInterrupted`). When the microsymbol is created the begin timestamp is set; the end timestamp is only set when no more object or person is detected in the sensitive area. If there happens to be more movement later on, a new microsymbol is created.

If a motion detector has time hysteresis, meaning that it triggers upon movement and stays active for a defined time period, this behavior is reflected in the microsymbol: it is created and the end timestamp is set when the hysteresis ends. The time representation of this symbol is "period".

Properties

`position`: The position for a motion detector is represented as a wedge or a cylinder, depending on the active area of the sensor.

`omsFootprint`: When a tactile floor sensor is triggered, this microsymbol is created. Along with all other microsymbols, a footprint does not carry along the physical properties of the sensor; in the case of the footprint, this means that the contact, which triggers it, has to be de-bounced in order to avoid creation of too many microsymbols in a short time. `omsFootprint` is created by tactile floor sensors; if these are replaced by more advanced sensors that are able to measure weight or dynamic characteristics of a walking person, this microsymbol will have to be extended (or replaced by another one). Since the sensors used here have only a binary output, they indicate merely whether they have been triggered or not. Similar to `omsInterrupted`, this microsymbol has a begin timestamp that is set upon creation and an end timestamp that is initialized with infinity and set when the sensor is no longer being triggered. The time representation of this symbol is "period".

Properties

`position`: The position for a tactile floor sensor is represented as a point or a rectangular, depending on the active area of the sensor.

`omsNoise`: The basic microsymbol that is created by a microphone. It indicates that the microphone has received sound with an intensity higher than a definable threshold. The microsymbol is created when the sound is received with the end timestamp set to infinity; the end timestamp is set when the sound falls below the threshold. This microsymbol is intended to contribute to person detection in a room, assuming that persons make noises. Of course, it will also be triggered by any machine or event that is noisy, but the decision whether a person is in the room is done at a higher level anyway. The time representation of this symbol is "period".

Properties

`soundPosition`: The position is the room where the sound is detected or, in case of big rooms, a rectangle that is limited to the vicinity of the sensor.

`omsSoundSource`: This microsymbol is created by a microphone array that is able to locate sound sources. If the underlying algorithm is able to detect multiple sound sources, then multiple microsymbols are created. The time representation of this symbol is "period".

Properties

`soundPosition`: The position of the sound source represented as a point.

11.4.3 Snapshot Symbols

This section contains the list of snapshot symbols that are created by the Person Tracking application service.

`osnPerson`: When the system has found evidence that it currently perceives a person, it creates this symbol and sets the property `position`. The information contained in this symbol is sufficient to create a representation symbol of a person, the representation has no access to microsymbols.

Properties

The following are the properties in `osnPerson` that are set by the application service Person Tracking:

`position`: The position of this person as it can be derived at this time. From the sensor information that is contained in the microsymbols of the Person Tracking application service, different possibilities arise. A single light barrier can only define the position of a person to be along a line, a motion detector has location information in the shape of a wedge or a cylinder. The determination of the position takes all available position information into account; this includes all microsymbols that contribute to position information and the quality of their information. The resulting position is in general a shape in three-dimensional space, where a few simplifications may be applied. Persons have a shape that cannot be perceived by any sensor (with the exception of cameras). This information is added to the snapshot symbols without actually having been registered by sensors. A light barrier, which is triggered by a person has no information about the z-axis of the position, therefore the snapshot symbol in this case will have a z-axis that is on the floor – assuming that persons remain on the floor most of the time. Only in special cases (e.g. the child safety application described in section 9.6), the z-axis position will be different from being on the floor.

The following microsymbols contribute to position information with a specific geometric shape (the sensitive area of the sensors triggering the microsymbol):

- light barrier: line
- motion detector wedge or cylinder
- floor sensor: point or rectangle
- microphone: rectangle
- microphone array: point
- height sensor: rectangle

Overlapping microsymbols are used to further determine position and result in other geometric shapes. A snapshot symbol always contains only one geometric shape as the position of a person, which is a point, a line, a rectangle, a triangle or a circle[53].

direction: the direction of a person as a three dimensional vector.

height: the height of the person as it originates from the height-measuring sensor or the set of light barriers

osnPersonEnters: When the system has found that a person leaves one room and enters another, this snapshot symbol is created. It is closely related to the definition of the office environment defined in section 6.1.1. Based on the different microsymbols, snPersonEnters indicates that a person has left one room and entered another one. Room refers to a real room, not to the cuboids that make up a room.

Properties

roomFrom, roomTo: the room identifiers of the two rooms that are involved.

---

[53] Whenever the output of an intersection cannot be mapped onto one of the basic shapes (e.g. intersection of a circle with a rectangle), simplifications are applied, i.e. the resulting shape is described by one of the basic shapes.

11.4.4 Representation Symbols

`orpPerson`: The application service Person Tracking creates this symbol when a new person is identified. In existing symbols, the Person Tracking application service sets only the property `position`. This symbol is based on the snapshot symbol `snPerson`; while snapshot symbols are created every time the system perceives something new (which can be multiple times per second), the representation symbol persists as long as the person is within the area that is observed by sensors. The task here is to connect single snapshots of persons to one continuous representation of a person.

Properties

The application service Person Tracking sets only one property in `orpPerson`:

`position`: The position as it is calculated by the representation module. It takes the position of the snapshot symbols `snPerson`, both present and past positions and finds the most reasonable position.

11.5 Application Service: Human Activities

The application service Human Activities adds additional information to a person, indicating which activity this person currently pursues. The activities, which are in the scope of the application service include: `actLie`, `actSit`, `actStand`, `actWalk`, `actRun`, `actTalk` as general purpose activities (see section 9.4) and `actWork`, `actMeeting` as higher-level activities that are set by Person Surveillance (section 11.6).

11.5.1 Activities

The different activities are based on various events and objects that need to be present. The activity `actLie` is set when the person is still and an object feasible for lying (i.e. a bed) is next to the person. Similarly, `actSit` requires a still person and a chair to be present, while the `actStand` activity only requires a still person without any additional objects. The activities `actWalk` and `actRun` are determined by a speed boundary between the two (another speed boundary is between `actStand` and `actWalk`). Determining whether a person talks would require speech recognition abilities or at least deep aural data analysis, therefore `actTalk` is simplified by defining that a person has the activity talk, when there is audible sound in the room detected by the microsymbol `omsNoise` or `omsSoundSource`. A person is considered to be working when a table and a chair are present (basic requirements for desktop work). The activity `actSit` is therefore a trigger for the activity `actWork` (see section 11.6). A meeting is defined as having at least two persons in a room and the activity `actTalk` occurs as well.

All the above described activities have a longer time horizon, meaning that, unlike the snapshot symbols, the time frame for activities is not an instant (or short period of time), but rather a period of some seconds or minutes.

11.5.2 Representation Symbols

The application service Human Activities does not create new symbols, but rather sets properties in the representation symbol `orpPerson`.

Properties

The activities described above are mapped directly onto properties of the representation symbol `oprPerson` with a prefix to indicate activities: `actLie`, `actSit`, `actStand`, `actWalk`, `actRun`, `actTalk`. These activities can partly occur simultaneously, some of them are required for higher-level activities that are set by Person Surveillance (section 11.6).

Additionally, Human Activities sets the property `lifeStage`, which classifies children, grown-ups and seniors. Children can be separated from grown-ups by analyzing visual data or by obtaining height information from an array of light barriers; the presence of a senior has to be available as a priori knowledge, because they are a group of people, who require special treatment, but cannot easily be distinguished from regular grown-ups.

11.6 Application: Person Surveillance

Person Surveillance relies on Person Tracking and Human Activities to determine the position of persons and analyzes the activities of persons based information. It is responsible for creating the representation symbols for persons and for updating some of their properties (see below). This application has two tasks: determine the position for person symbols on representation level and create additional activities (derived from basic activities of Human Activities).

Aside of the central representation symbol `orpPerson` the application Person Surveillance does not create any other symbols.

`orpPerson`: Like other representation symbols, this symbol is rarely created or destroyed, but frequently updated. If such a symbol shall be created (e.g. because a new person enters the environment), then it is done by Person Surveillance. By evaluating snapshot symbols (`osnPerson`), the property `position` is frequently updated; the two higher-level activities `actWork` and `actMeeting` are set.

11.7 Application: Child Safety

Child safety observes persons that are classified as children and detects possibly dangerous scenarios. Aside of knowing the persons position, the possibly dangerous objects need to be considered. These objects come from Object Tracking (section

11.2) and are table and stove. Regarding tables, the system has to observe children that climb on top of it and regarding the stove, the possibly dangerous situation is an unattended child near a hot stove. Both are treated as scenarios, therefore the system sets scenario symbols, when it detects one of the situations.

The application Child Safety does not create or modify any symbols on representation level or below, it only creates scenario symbols and action series symbols.

11.7.1 Scenario Symbols

`oscChildInDanger`: When a possibly dangerous scenario has been detected, the application Child Safety creates this symbol.

Properties

As additional information, the symbol `oscChildInDanger` contains the property `dangerType`, which indicates the type of possibly dangerous situation that has been detected (which can be either the hot stove or the child climbing on a table). Furthermore, there is a link to the child, which is possibly endangered; this is stored in the property `person`. If there are objects involved (which is the case in the two sample scenarios covered here), a link to these objects is also available in the property `objects`.

11.7.2 Action Series and Action Symbols

`oasInformOperator`: A reaction to the detected scenario is the notification of a human operator. As with all action series symbols, `oasInformOperator` is processed by the Action Subsystem (section 11.10).

Properties

When triggering the Action Subsystem, the necessary information for the system to react accordingly needs to be present in the action series symbol. The information that Child Safety passes on when requesting a notification to the operator, contains the cause (stored in property `cause`, which contains the information that a child is in danger) and the persons and objects involved (stored in the properties `person` and `objects`).

11.8 Application: Geriatric Care

Geriatric Care supports supervision of elderly people by three tasks: it detects possibly dangerous situations, which have been defined as a senior collapsing and a senior staying in a cold room for long time (section 9.7), it helps to locate seniors who are disoriented and cannot be found by the supervising personnel, and it helps the senior by locating objects of interest (e.g. glasses, keys or books – which, for this work, have been substituted by a placeholder object).

11.8.1 Scenario Symbols

`oscSeniorInDanger`: When a possibly dangerous scenario has been detected, the application Geriatric Care creates this symbol.

Properties

The symbol `oscSeniorInDanger` contains the property `dangerType`, which indicates the type of possibly dangerous situations that has been detected. The property `person` is used for providing a link to the senior in question.

11.8.2 Action Series and Action Symbols

`oasInformOperator`: A reaction to the detected scenario is the notification of a human operator. As with all action series symbols, `oasInformOperator` is processed by the Action Subsystem (section 11.10).

Properties

Geriatric Care passes the action series symbol `oasInformOperator` to the Action Subsystem when requesting a notification to the operator; this symbol contains the cause of the notification (stored in property `cause`) and the persons involved (stored in the property `person`).

11.9 Application: Comfort

Climate and lighting control is the goal of the prototypic Comfort application. It has sensors to gather information about the climate (temperature and humidity) and actuators to control climate and lighting. Any changes in comfort are redirected to the Action Subsystem, which processes the action series symbols.

11.9.1 Microsymbols

`omsTemperature`, `omsHumidity`, `omsBrightness`: These microsymbols provide the information about the climate and lighting level in a room. Each of these symbols can be created by a single sensor or by a set of sensors. On microsymbol level the information is provided, but not processed further, this is done in the according snapshot symbols.

11.9.2 Snapshot Symbols

`osnRoom`: This symbol is used to make the microsymbols `omsTemperature`, `omsHumidity` and `omsBrightness` available on presentation level. `osnRoom` has three properties `temperature`, `humidity` and `brightness` that are used to represent the distribution of the according physical variables over the room. In a room with only one sensor, this information is taken directly from the sensor; in rooms with multiple sensors, the information of the scattered sensors is used to represent the distribution.

11.9.3 Representation Symbols

`orpRoom`: Taken from the information of `osnRoom` to provide a representation level symbol for rooms. The properties `temperature`, `humidity` and `brightness` represent the distribution of temperature, humidity or brightness, respectively. The room temperature is used in the Geriatric Care application to detect a possibly dangerous scenario.

11.9.4 Action Series Symbols

`oasClimate`: Depending on the information available about temperature and humidity, and supported by information from Person Surveillance, the application will issue an action series symbol to alter the climate. The information that is associated with this symbol depends on the implementation of the system actually controlling temperature and humidity.

`oasLighting`: Upon presence of persons in a room, the system controls the brightness level of the according room. On the other hand, energy management purposes cause the light to be dimmed or turned off. The according scene[54] is contained in the action series symbol `oasLighting`; the information that is associated with this symbol depends on the implementation of the actual lighting control system.

11.10 Action Subsystem

The Action Subsystem processes action series symbols and executes the required actions. An executed action is added to the representation by introducing a new action symbol. Since the ARS system is mainly observing and not strongly interacting with its users, the main action series symbol is `oasInformOperator`, which causes the execution of a single action symbol, `oacInformHumanOperator`. Additionally, there are action series symbols for the Comfort application: `oasClimate` and `oasLighting`. The execution of both depends on the lighting and comfort system that cooperates with the ARS system and is not included into the symbolic description in this work.

11.11 Inner World

The symbols that are part of the inner world are used for maintenance information and status of the system components. Like in the outer world, there are different levels of symbols: microsymbols, snapshot symbols and representation symbols are used for

---

[54] "Scene" in this context is not related to the symbolic "scenario". It refers to the setup of light in a room and its control.

creating the world model; action series and action symbols are used for activities that the system initiates.

The amount of symbols is limited, compared to the outer world and represents only a limited model of the system and its modules. The focus of this work lies in the perception of the outer world, therefore the inner world does not contain as many symbols as the outer world representation. At the current state of technology, it appears sensitive not to push the analogy between a biological system and a technical system too far. Since the ARS system is not driven by its needs in a way a living being is, the inner world is reduced to maintenance purposes and status information.

11.11.1 Microsymbols

imsSensorFailed: A sensor with self-testing ability causes this microsymbol to be created. It indicates that the sensor is currently not operating and needs maintenance.

imsActuatorFailed: Like imsSensorFailed, this microsymbol indicates a faulty actuator.

imsConnectionLost: if components are unable to detect failures, it is still possible to detect that they are not available. Communication requests that remain unanswered by a component cause the creation of this microsymbol. The cause for a missing reply may be either a faulty component or an interruption of the communication channel. An imsConnectionLost symbol can therefore also indicate that more than one component is not reachable.

11.11.2 Representation Symbols

irpSensor: The symbolic representation of a sensor that is part of the system. It has two properties, position and status. These are used to determine the position of the sensor, i.e. where it is mounted and its status, indicating whether the sensor is operating correctly.

irpActuator: Similar to irpSensor this symbol represents an actuator and has the properties position and status.

irpComponent: This symbol represents a general component that executes some part of the system code (e.g. desktop computer or embedded system) and does not have sensors or actuators. It has two properties, position and status. These are used to determine the position of the component and its current status.

irpLink: Describes the communication link between components. It has a property status that indicates correct operation and a property components that contains all components, which are connected using this link. status depends on the microsymbol imsConnectionLost.

11.11.3 Action Series Symbols

`iasInformOperator`: If the inner world application has detected a failure or provides a status update of the system, it creates an `iasInformOperator` symbol containing the cause for notification (property `cause`) and the components that this notification refers to (property `components`).

11.12 Symbol Overview

The following tables are a compilation of all symbols that are defined in this work. They give an overview of the connections between application (and application services) and the symbols that belong to them. Table 11.1 and Table 11.2 are tables for outer world modules. Table 11.1 shows, which module creates which symbol. If there are properties that are set upon creation, they are mentioned in the second column.

Symbol	Property	created by
`omsInterrupted`	position	Person Tracking
`omsObstructedLb`	position	Person Tracking
`omsObjectPassesByLb2`	position, direction	Person Tracking
`omsPersonHeightIrH`	position, height	Person Tracking
`omsObjectMovesMd`	position, height	Person Tracking
`omsInterrupted`	position	Person Tracking
`omsFootprint`	position	Person Tracking
`omsNoise`	soundPosition	Person Tracking
`omsSoundSource`	soundPosition	Person Tracking
`omsTemperature`		Comfort
`omsHumidity`		Comfort
`omsBrightness`		Comfort
`osnPerson`	position, direction, height	Person Tracking
`osnPersonEnters`	roomFrom, roomTo	Person Tracking
`osnTable`	position, dimension, height	Object Recognition
`osnChair`	position, seatHeight,	Object Recognition

Symbol	Property	
	chairBack	
osnOrangeBall	position	Object Recognition
osnRoom		Comfort
orpRoom		Environment Model (upon start-up)
orpTable	position, dimension, height	Object Tracking
orpChair	position, seatHeight, chairBack	Object Tracking
orpOrangeBall	position	Object Tracking
orpPerson	position	Person Tracking
oscChildInDanger	dangerType, person, objects	Child Safety
oscSeniorInDanger	dangerType, person	Geriatric Care
oasInformOperator	cause, person, objects	Child Safety, Geriatric Care
oasClimate		Comfort
oasLighting		Comfort

Table 11.1: Outer world symbols and their properties are created by the module in the third column

Table 11.2 shows the properties for outer world symbols that are not set upon creation, but can be modified or updated. An update does not necessarily originate from the application that created the symbol.

Symbol	Property	modified by
orpPerson	actLie, actSit, actStand, actWalk, actRun, actTalk	Human Activities
orpPerson	actWork, actMeeting	Person Surveillance
orpPerson	lifeStage	Human Activities
orpOrangeBall	possessor	Person Surveillance
orpRoom	temperature, humidity, lighting	Comfort

Table 11.2 The module in the third column modifies properties of a symbol

Table 11.3 gives an overview of the symbols used in the inner world. Since there is only one application responsible for the inner world, the table is for both creation and updates of symbols and their properties.

Symbol	Property	modified by
`irpSensor`	`position, status`	Inner World
`irpActuator`	`position, status`	Inner World
`irpComponent`	`position, status`	Inner World
`irpLink`	`status, components`	Inner World
`iasInformOperator`	`cause, components`	Inner World

Table 11.3: All inner world symbols and properties are modified by the inner world application

12 Communications Design

The system has a strong need for communication. Modules need to interchange symbols and property updates, sensor data need to be processed and visualization needs to be informed of changes in the graphical representation of symbols. Under the assumption that the system shall be able to process lots of sensor information, an analysis on how to provide a proper communication system for this data is done in the following sections. Aside of field level issues, the link between office networks and fieldbus system is also examined.

12.1 Integration between Field Level and WAN

Networks have penetrated both the domain of automation and the office world, a development, which today provides us with vast possibilities for interconnection of field level devices and high-level IT logic. The reason for these advances can for once be seen in the success of the Internet, which is today omnipresent and builds the base for many different applications. On the other side fieldbusses have sufficiently evolved and been properly standardized in the last years, so that they have become reliable communication systems for building automation. LonWorks nodes, for example, can be programmed like compact embedded systems using Neuron C [Ech95] – a derivate of the C programming language. Today the quest for an integration of both worlds – the fieldbus world and the office world – has culminated in what is commonly called *vertical integration*. Systems like the ARS system, but also a variety of different applications ranging from Enterprise Resource Planning (ERP) to large-scale energy management require data to be available at a level, which is outside the domain of fieldbusses. Vertical integration stands for seamless integration of information over all levels of an enterprise or data processing systems, an inclusion of data from the field level into management applications; it also considers the other direction, providing direct access from the management level down to the process control[55].

Horizontal integration, on the other hand, looks at integration on the same hierarchical level. The communication technology base of horizontal integration is essentially the network designed for each application. The company or office level is dominated by TCP/IP [RFC793] based networks running on Local Area Networks (LANs), which and are therefore the de facto standard at this level. TCP/IP has the additional benefit of providing seamless connection to Wide Area Networks (WANs).

On the field level, fieldbus systems are established; development today focuses on integration of alternative systems, like, for example, wireless technologies, mobile nodes or self-contained wireless sensors that do not need external power supply. Ethernet is a candidate to enrich the palette of fieldbus system, Industrial Ethernet is determined to replace existing systems, but still has to overcome issues like determinism or real-time issues.

A great step towards horizontal integration has been achieved by moving away from defining only the lower layers of communication, which are responsible for reliable data transmission, but to introduce profiles, which are located on application layer and above. These allow interoperability between devices of different vendors not merely on the level of matching plugs and communication protocols, but in terms of correct semantic interpretation of data that is exchanged between the devices. Lately this process was extended to use the Extensible Markup Language (XML) as a common basis for describing devices and their functionality.

Approaches for Linking the two Domains

In order to provide a connection between the field level and the office level applications there are different architectures available. It has to be decided where to cut one system open and "glue" it to the other. The main goal that the ARS system looks at in this field, is how to overcome the differences between the various different fieldbus systems and how to spatially separate field level installations from higher-level processing of fieldbus data. This results in the task to connect fieldbusses to IP-based networks, where two general classes of interlinking are possible: either protocols are tunneled, thus bridging complete segments and virtually integrating one system into the other; or the cut is made on application level using a *gateway* [Pra01] that provides more or less abstract high level services for fieldbus interaction. Apparently, it is not possible to join networks of the two different domains into one

---

[55] The term vertical integration actually originated from the domain of economics and management. There means that a company is engaged (in the sense of ownership) in several stages of a given industry's value or supply chain [Rud03].

global network; therefore, it is always necessary to have a device in the middle that deals with these connections, which shall be called gateway. The reader shall be aware that the term gateway is used with many different meanings, depending on the domain one works in, and is therefore a bit overstressed. Unfortunately other alternatives (e.g. access point) are similarly overloaded; therefore, the term is used in this work anyway with its exact meaning being defined in the following.

Tunneling

The tunneling approach shows two alternatives: either the fieldbus protocol is tunneled over the Internet Protocol or vice versa. In any case, the data packets from one protocol are taken and wrapped as opaque payload data into packets of the other protocol (with all implications that this behavior implies). In cases where protocols of the same protocol suite are tunneled (e.g. an HTTP tunnel to overcome restrictions by firewalls), it is possible to decide about the layer, where tunneling shall be established – which need not necessarily be the data link layer[56]. However, when connecting networks that are as different as fieldbusses and IP networks, the most feasible solution is to tunnel complete link layer packets.

A requirement of tunneling is that both communication partners understand the same protocol. The interconnecting network is merely used for opaque data transport and shall at best not expose any properties that could affect the connection between the communication partners (latency, for example, need to be within the constraints of the tunneled protocol). As soon as a complete tunneled Protocol Data Unit (PDU) reaches its destination, the application on this side has to be able to understand the syntax and semantics of the protocol being used.

Tunneling a fieldbus protocol over an IP network is used when interconnecting fieldbus segments over longer distances. The physical media that can be used for a fieldbus are usually limited in length, which is needed for the design of timeouts and other parameters. Therefore, if it is necessary to connect a remote fieldbus segment, a tunneling approach can be chosen. While bandwidth is usually not a concern, the latencies that are introduced by the tunnel may affect the functionality of the remote fieldbus segment, especially if real-time is a requirement. Unless a backbone is available, which is able to provide the required quality of service, the tunneled connection has to be handled in a special way and cannot seamlessly integrate into the other fieldbus segments.

If real-time is not a requirement and the remote connection is sufficiently fast (meaning low enough latency times), the IP tunnel allows for a cost effective connection between remote installations. A feature that greatly simplifies

---

[56] Using the lowest possible layer for tunneling has the advantage of not losing information; on the other hand, a lot of information is useless and still needs to be transmitted, for example, checksums of the original protocol, which are also treated as application data.

implementation of this kind of interconnection is the ability of a fieldbus to support bridges between fieldbus segments. This means that the fieldbus nodes are aware that some nodes are not reachable directly, but have to be handled specially (e.g. by setting longer timeouts).

The second tunneling solution is to encapsulate protocols from the IP-suite into fieldbus protocol packets. Today there are devices available that implement a TCP/IP stack together with a small web server as a single chip solution. An application can therefore directly contact a node using an HTTP [RFC1945] connection and retrieve information in XML format. Although an understandable requirement that provides convenient vertical integration, the implementation of a web server on each node has some downsides. For once, the overhead for tunneling HTTP using the (usually limited) bandwidth of a fieldbus, is considerable. Second, the application that wishes to retrieve data has to deal with a considerable amount of web servers on the field network[57]. Seen from this point, a web server should not be located at each node, but rather be a central instance that provides data from the fieldbus nodes (which goes in the direction of a gateway as described in the next section).

Another downside of tunneling IP over a fieldbus protocol is addressing of fieldbus nodes. Aside of the native fieldbus address, every node that shall be connected with the IP network needs to have its own IP address. Private networks with their own pool of IP addresses are an option to overcome problems with the limited address space of the currently dominant IPv4. This would however restrict access to the nodes and requires means for offering server services (like the above mentioned web server) to hosts on the public Internet.

Finally, two more reasons make this tunneling approach unfeasible: first, fieldbus protocols are designed for low latencies to provide timely transmission. IP packets can – depending on the underlying medium – be considerably larger than fieldbus packets. This requires the packets to be fragmented into rather small pieces. This for once creates a lot of packets on the fieldbus and secondly introduces even more overhead, since the fragmentation into small packets is a costly operation (costly in terms of network bandwidth). The second problem occurs in fieldbusses that do not offer peer-to-peer communication between the nodes, but rather master-slave type of communication. In IP networks, every participant is able to communicate with any other node of which it knows the address. It is possible to implement peer-to-peer communication using master-slave communication, but this introduces considerable delays, if a slave wishes to send by itself. It has to wait, until it is polled by the master. Furthermore, some protocols of the IP protocol suite require broadcasts over the subnet, which is especially hard to implement on a master slave system.

---

[57] It appears more reasonable to implement another protocol, for example, the *Simple Network Management Protocol* (SNMP) [RFC1157] or the *Lightweight Directory Access Protocol* (LDAP), which uses less network bandwidth.

Gateway

The previous section has shown that tunneling of protocols has a number of disadvantages, therefore this sections looks at a different approach: instead of wrapping protocols into packets without understanding the contents of the PDUs, the gateway introduced here is capable of understanding protocols on both sides and acts as a translator between the two domains. The gateway is for once a node on the fieldbus side, and it can be accessed through IP-based mechanisms via the Internet. What is different compared to the tunnelling approach is that a client does no longer directly connect to a server that runs on a fieldbus node. Instead, the gateway represents the fieldbus and its data to the IP-network. Upon request, it fetches the data from the fieldbus nodes by native fieldbus communication means[58]. It has to be stated that real-time requirements are not considered in this approach; the gateway terminates protocols on both sides and thus the according quality of service properties are not available. Since hard real-time is usually not a requirement in building automation (with the possible exception of safety critical applications) and is definitely not a requirement of the ARS system, it can safely be dropped here.

The architecture of such a gateway can be done in different ways. The issues of data availability and fieldbus data representation are covered in section 12.2. Two basic possibilities arise when designing such a gateway: it can be designed to support a single type of fieldbus. In this case it should seek to support all features of this fieldbus and find the best possible way to seamlessly transport the properties and services of the fieldbus over the IP-based network. Or, as a second approach, the gateway is designed to support a lot of different fieldbusses. In this case it cannot provide every feature of every possible fieldbus at its Internet front-end, but should rather seek to find a subset of features that it can provide for all supported fieldbusses. This topic is tackled in more detail in section 12.2. The goal of the ARS system is in any case not to be dependent on one specific fieldbus and be able to obtain data from different fieldbus systems, therefore the gateway, which is able to support multiple fieldbusses, is the architecture of choice.

---

[58] Or, as an additional feature, the gateway is able to store a *process image* of current and consistent data that it can fall back on.

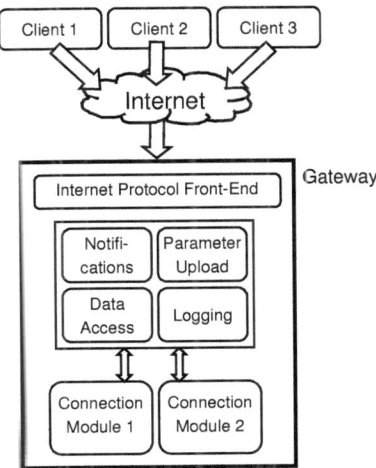

Figure 12.1 Gateway architecture for abstract gateway services

As for the Internet side of the gateway, a lot of research has been done and published. In [Lob02] an overview of different Internet front-end protocols is given together with an analysis of their feasibility. The architecture of the gateway that is described in [Lob05] is shown in Figure 12.1. Clients connect to the gateway using the Internet Protocol Front-End; the core of the gateway are the services, which provide data access and logging of data as well as the possibility to update device parameters and create notifications that depend on the data. Logging is important for gateway connections that are either costly (e.g. dial-up lines) or do not provide a reliable communication channel, e.g. power-line communication [Lob03]. Fieldbus access is provided by different connection modules, where each module is able to connect to one type of fieldbus. Standard protocols like SNMP (Simple Network Management Protocol) [RFC1157] or web technologies like HTTP are in terms of feasibility beaten by LDAP [RFC2251] or a proprietarily defined protocol. The gateway in any case provides an abstract interface to the fieldbus systems, independent of both the fieldbus protocol and the fieldbus-specific coding of the data, which enables it to provide general purpose access.

12.2 Fieldbus Data Representation and Gathering

The previous section identified a gateway capable of supporting multiple fieldbusses as a possibility for the ARS system to gather data from different fieldbusses. This section describes the requirements for fieldbus interconnection in more detail. The gateway has to implement a set of services; it has to provide an Internet front-end using a specific protocol and it needs to have a representation of fieldbus data that is able to cover the supported fieldbusses.

The gateway discussed here has an architecture that attempts to find an optimum between fieldbus dependency, implementational effort, and data abstraction [Saut02]. It supports different fieldbusses by implementing different fieldbus connection modules, it has a central processing part which is responsible for routing manipulation requests to the fieldbus, scheduling and processing events and providing access to data points and it offers an Internet front-end (see Figure 12.1). The interfaces between these three modules are by themselves all based on TCP/IP, thus not only providing clean and well-defined communication between the components, but also allowing to distribute these modules themselves across a network. This can be useful if not all fieldbusses are directly attached to the gateway, but are only accessible via an additional IP-connection. Because the fieldbus modules communicate only using a TCP/IP-based protocol, it does not matter whether they run on a remote machine, as long as it has a network connection.

A central concept of the gateway is the *data point*. The fieldbus network is represented as a set of nodes, each uniquely addressable. Every node has a set of data points, also uniquely addressable, which represent values that a node is able to provide upon request. Most of the data points are scalar, but types that are more complex are also possible. An important feature of data points that originates from fieldbus nodes is the timestamp: a data point carries a timestamp that is set when the current value is set[59]. This timestamp stays attached to the value of the data point all the way up through the processing application. This way the timing stays appropriate, even if transmission or processing of the value should be delayed.

The gateway offers the following services: it allows reading data points as well as writing of data points (which is required for control applications as well as for configuration purposes) and it is capable of storing historical data that can be conveniently retrieved in a *log*. This feature is not relevant for the ARS system, since it stores data in a central database, but is convenient, when the IP-network is not permanently available, but rather implemented as a dial-up connection[60].

Fieldbus data representation is a complex task that cannot be solved sufficiently for every existing fieldbus. Depending on how detailed the data representation is modeled (and required by an application) the representations can strongly vary. A temperature sensor, for example, can be modeled as providing a room temperature in degrees Celsius, which would be one scalar value. However, additional information about this temperature might be relevant, for example, its accuracy or the time it takes until the sensor can reliably react upon changes in temperature. Even if these issues are

---

[59] Timestamps are further discussed in section 12.5.

[60] Another feature that is related to this and is also not used is the ability to send notifications. The gateway can monitor a preconfigured value and raise a notification, if a criterion is met. This notification can be implemented within a protocol like SNMP or it can be a proprietary implementation (e.g. sending an email)

resolved, the question of numerical representation remains. How many digits of precision are required and in which binary form shall they be represented? The decision, whether to use integer numbers (which are fast to process), fixed point numbers or floating point numbers, has to be made. Definitions like "this integer number represents a temperature where 0 represents 300°C and an increase by one means increasing the temperature by 0.1°C" are not uncommon and need to be taken into consideration when finding a generic data representation.

The data point concept therefore describes the representation of values and attempts to find a tradeoff that suits the requirements. A data point as described here contains not only a value, but also has the following attributes: type, encoding and access mode. The *type* tells the system how to handle the value in the data point. Every data point has a certain type, e.g *integer*, *double* or *string*. A special data point type is *raw*, which is an opaque type for data that does not fit anywhere else. This especially applies to fieldbus data that are available only in form of a structure and not as a scalar. The *encoding* is a rule for the representation of a data point value. Different encodings are available, for instance *string*, *number* and *time*. Typically, a data point can be represented using more than one encoding, but not every encoding is valid for every data point. The *raw* encoding is available for any kind of data point value. It provides a byte-level view of the value and is therefore somewhat dependent on implementation details. The *string* encoding is used almost exclusively for data points of type string, whereas number and time encoding are numerical encodings. The *access mode* describes what a client is allowed to do with a data point, namely whether it may read and write a data point or only read a data point.

12.3 Communication Framework

The system prototype consists of many different modules that are designed to be easily distributable over several computers. To achieve this distribution, the communication framework that connects the modules has to provide the necessary services for communication. In the course of the ARS project, *SymbolNet* has been designed as the framework that shall connect all components of the system, which are related to symbolic processing of data. It is implemented in Java for reasons of built-in security in the language, good support for networked debugging and short deployment times; additionally, parts of the implementation are also available in the programming language C. SymbolNet allows application level data exchange and synchronization. Its main task is to communicate creation, destruction and updates of symbols and their properties between the modules. Updates occur transparently for the modules; this means that an application can modify a symbol property and the change of this property is communicated by the framework directly to all modules that have announced interest in the according symbol. Accordingly, an application receives an update of a property and can process it. Since the system has many powerful components (microsymbol factory, perception module and representation module as well as the simulator or the sensor database), it has to be distributed amongst different machines, thus requiring a network transparent design.

Aside of better scalability of modules when using a distributed, network transparent architecture, SymbolNet also enables debugging and communication load analysis, by disclosing the communication interfaces between modules. The modules receive symbols, process them and issue new symbols (or symbol updates). All this communication can be logged in a time consistent manner to see the input and output of a module and detect bottlenecks.

The concept of SymbolNet foresees a graph-like communication structure between the modules as shown in Figure 12.2. Every module can be the originator of symbols, which it fills into a pipeline. At the end of the pipeline, there may be one or more modules that receive the symbol.

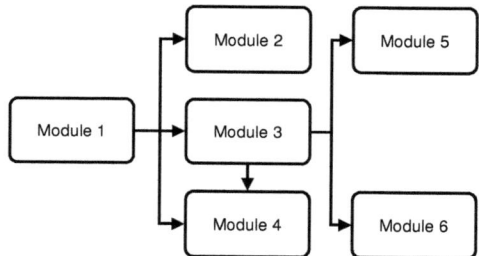

Figure 12.2 Communication structure of SymbolNet

12.3.1 Message Format

The protocol that is used to communicate changes in the state of symbols (creation, termination, updates) uses messages that are built in *Abstract Syntax Notation number One* (ASN.1) [ASN02], because this allows for easy extension of the protocol format and contents. ASN.1 was developed to provide an abstract syntax, which can be used for exchange of data between communication systems. It is used by standardization organizations in combination with textual system descriptions as well as by several formal techniques to describe data and the structure of protocol data units (PDUs). Using ASN.1 it is possible to describe data structures and interfaces in the upper OSI protocol layers. ASN.1 is associated with *coding schemes*, used to create binary representation of the specified PDUs; the encoding scheme used for SymbolNet is DER (distinguished encoding rules [DER02]). ASN.1 does not define formal semantics or operations to manipulate data; this has to be done in a separate protocol definition. There are three core messages: creation of a new symbol, destruction of a symbol and update of symbol properties. A module that receives a message can decide if it wants to process it or not. In any case, it may have to propagate the message to any other receivers that are connected to the module.

12.3.2 Communication Infrastructure

The core communication infrastructure for symbolic modules is shown in Figure 10.1 in chapter 10. Although exchange of symbols is done hierarchically from microsymbols over snapshot symbols to representation symbols, the infrastructure allows every module to communicate with every other module. This is necessary, because higher-level symbols (on representation level and above) may be used and modified by different modules without having a dedicated hierarchy. SymbolNet is always backed up by a database (section 12.3.3), which guarantees consistent data access.

At the borders of symbolic processing there are on one side sensor data and on the other side simulation and visualization. Because of the Fieldbus-Internet gateway described in section 12.1 different fieldbusses can be integrated into the system with only one interface to be adapted. Additionally, it is possible to circumvent the gateway and use a native connection to sensors and actuators by directly accessing the sensor database (section 12.4.1).

To visualize both simulated and real data, envSim (section 13.1) is connected to the communication framework. By using a reduced version of SymbolNet, envSim gets updates about the currently active symbols and their properties. The graphical representation of the symbols is preprocessed, so that envSim merely has to overlay graphical representations over the rendered environment.

12.3.3 Database Access

SymbolNet is the communication framework that provides efficient and convenient information exchange between the components of the system. It emerged from the need to have a consistent database between all participants. A first solution envisioned a database to be the central information sink, where each component would deposit and retrieve information[61]. A purely database-centered infrastructure is however not feasible, because of two reasons: when a symbol is created by one module and communicated to another module, the communication path would consist of writing into the database and reading out of it. This would produce considerable propagation delay between the modules. Second, a requirement is to use only non-proprietary database access mechanisms in order to avoid dependencies on one specific database. Thus, a module would have to poll the database permanently for updates of relevant symbols. This would cause considerable performance issues, assuming that all modules in the system would have to poll the database.

The solution to circumvent possible bottlenecks is the introduction of SymbolNet. It reacts event-driven, meaning that messages are issued upon creation or updates of

---

[61] The symbol database is responsible for storing all types of symbols from microsymbols to action series symbols, while the sensor database stores only sensor values. This separation is necessary due to the different nature of the data and the various means of data access.

information. Still, the consistent set of data has to be preserved. Therefore, SymbolNet is the component that keeps contact to the database and communicates all updates to the database. Whenever a symbol is created, destroyed or updated, SymbolNet transparently writes the new information into the database. This way, the historic data is available in the database, which gives an additional possibility: for simulation and visualization purposes, the acquired data can be reused later. If real-time processing of data becomes unfeasible (because of complex calculations), the simulation can be run without visualization; when simulation is done, visualization can be started and display the already generated information. Since sensor information is also stored (in the sensor database, section 12.4.1) this information can be used to do multiple runs over the same data, which is a useful feature for tuning module behavior.

12.4 Database Storage

The ARS system prototype consists of distributed module that can be run on different networked machines. Since the whole system operates on the same set of data, it is necessary to maintain consistent data during operation, which is done by using a database as the central storage component. As described in section 12.3.3, direct access to a database is not feasible for performance reasons, therefore SymbolNet has been introduced. The database is a relational database, which is used to store sensor and symbol data (section 12.4.1 and 12.4.2, respectively). When referring to two different databases, this actually means one database installation with two different table spaces. In order to stay independent of a specific database, the access to the data is done by using SQL (Structured Query Language) [Ach00] statements and communicating them to the database. The only component accessing the symbol database from within the system is SymbolNet; the sensor database is accessed by embedded hardware that is responsible for writing sensor values.

12.4.1 Sensor Database Description

The database scheme for storing sensor data is built around data points. A data point is a source of information and can have various properties. The data can be separated in two categories: static information and dynamic data. Static information contains all configuration information including description of data points, their location, type and quality of data. The structure of a data point is also described as static data. Dynamic data on the other hand contains the actual values that sensors create. While static data can be complex (containing, for example, human readable descriptions of a data point), dynamic data are compact, because they are frequently updated. Dynamic data is stripped from all redundant information and is optimized towards size (while still keeping the link to additional, static information). This way it is possible to store big amounts of data efficiently without redundancy, while at the same time keeping consistent meta-information about the data points.

12.4.2 Symbol Database Description

Symbols and their properties are stored in the symbol database, a database, which is logically separated from the sensor database. Similar to the sensor database, static and dynamic information is both available, but separated. On lower levels of symbolization, symbols have only a few properties (or none at all); at representation level, symbols generally have a lot of properties – but the total number of symbols is low. In order to store all different symbols in the same database scheme, the symbols table is separated from the property table. Symbols are linked to their properties by reference IDs; this way a symbol with all its properties can be conveniently retrieved. Since the symbolic alphabet is manually engineered (and not created by the system itself), both symbols and properties can only have predefined classes. Symbols have a certain lifetime, where microsymbols commonly only exist in one instance (or a short time period), while representation symbols have a long lifetime. During creation, the creation timestamp is set and the end timestamp is set to infinite. Properties have values and these values can change, therefore properties also have a begin timestamp. The end of a property value is given either by a new value (e.g. the position of the person is updated) or by the termination of the symbol. Therefore, they only have one timestamp.

Symbols and properties have unique identifiers and a referrer to the class they belong to (which is also unique). Static information for symbols and properties can be retrieved by using the class referrer. Since every symbol belongs to one class, the description of the structure is stored in the class description. While structure descriptions can be complex, the values of properties are stored in a compact way, consisting of property identifier, timestamp and value. This allows storing large amounts of data without redundant data descriptions.

12.5 Time Representation

Modeling of time is an important issue when designing the system. Different issues have to be considered when planning time distribution, representation and storage. The following briefly covers the concept of time as needed for sensor values as well as for symbols and describes how time is stored and how operations on time are done when representing it as a number. Since the system is distributed over different computers that are connected by a network, the issue of time synchronization between the modules running on different machines is also briefly tackled.

Aside of the "physical" definition of time, which deals with representation and calculation of time, there is another way of dealing with time and other physical variables in terms of symbolic representation; this sort of high-level symbolic time is described in section 4.2 and is not strongly related to time as discussed here. Since this part of the thesis describes implementational issues, symbolic time representation is not covered here, these sections describe time, as it is needed to build the system.

12.5.1 Instant and Time Period

Symbols have two concepts of time: an *instant* and a *time period*. The instant is an abstract representation; it is used for events that have a duration, which is short compared to the required system timing constraints. An example is measuring the height of a person as described in section 11.2. The underlying physical effects and the resulting sensor output are such that a set of light barriers is interrupted, possibly at slightly different times. If this happens in a reasonably short period of time, this time period is modeled as a single instant, meaning that although the process of interruption actually takes a finite time to finish, the modules, which process information above the level of microsymbols only see it as a single instant. In the data representation of symbols, the instant is also referred to as the *timestamp* of the symbol. If a symbol has a time representation of "instant", the timestamp for the beginning and the end of the symbol lifetime are identical.

A time period is defined as the time between the instant when it starts and the instant when it stops. This is directly reflected in the time information of symbols: symbols that need a time period (and not an instant), have a beginning timestamp and ending timestamp. While symbols always need to have a valid beginning timestamp, which is current time or earlier, the ending timestamp can be infinite. This is the case when a symbol is created without a defined time of termination.

Time synchronization is an important issue in the system. Sensors create data at a certain instant and the sequence when this data is created is important for the system. Take, for example, two light barriers that are used to detect the direction of a moving person. Both light barriers need to be properly synchronized in order to provide correct information when each of them has been interrupted. The timestamp for both events has to be created directly at sensor level (see also section 12.2) and not when written into the database or when received and processed by a module, because otherwise the delay and especially the delay jitter would spoil the necessary precision.

12.5.2 Time as a Number

Time that manifests in timestamps needs to be stored and processed by the system. The requirements for time are identified as follows:
- The resolution shall be below one second; a resolution of one tenth of a second is sufficient, higher resolutions are acceptable
- Handling of time has to be convenient and resource efficient, basic operations like adding and subtracting shall be fast
- Portability between platforms, programming languages and database manufacturers
- Where possible, storage of time shall follow existing standards or de facto standards
- Linear and monotonous time representation

There exists a standard for time representation, the ISO 8601 standard, the third edition being available since 2004 [ISO8601]. The standard establishes a moment in time as precisely or generally as the user or system requires it to be. It does so by using a string of characters, ordered from the largest representations to the smallest, with the smallest being seconds or fractions of seconds. Although standardized and fulfilling precision requirements, ISO 8601 has comparably high computational costs. Time is stored as characters, which need to be parsed, before they can be processed. Compared to storing time as a number, this parsing is a costly process and does not circumvent the problem of internal storage: after parsing, the time has again to be stored in a format, which allows performing operations on it.

Since the system stores data that represents symbols or sensor information in a database, another approach is to examine SQL (Structured Query Language) data types. SQL was adopted as a standard by the ANSI (American National Standards Institute) in 1986 and ISO (International Organization for Standardization) in 1987. It defines a date format that may be used; however, this date format has two disadvantages: for once the resolution is one second, which is insufficient for the purposes of the system. Second, although SQL is standardized, it is possible that commercial implementations do not support basic features of the standard – such as the DATE data type – and instead use some variant of their own. Oracle, a major database product provider, offers such a proprietary data type, which is called TIMESTAMP. It allows a precision of up to nine digits of fractional seconds, which would be sufficient for the purposes of the system, but would bind future development to Oracle databases.

Linearity is an issue that greatly simplifies code design. A commonly required operation is to calculate a time period, by subtracting the beginning instant from the ending instant. Although time itself is always linear and monotonous (at least in the non-relativistic context that is assumed here), the representation of time is not necessarily so. Daylight saving time, for example, causes glitches in the progress of time, by either skipping an hour or repeating an hour. Figure 12.3 shows October 30th, 2005, when summer time is changed to winter time by repeating the hour between 2:00 and 3:00. Although physical time continues (the y-axis indicating minutes since midnight of that day), the representation of time is ambiguous in the hour between 2:00 and 3:00 and would therefore have to be augmented by an additional flag that indicates whether the time is in summer or winter time. Otherwise, the time "2:30" can either represent 150 minutes after midnight or 210 minutes after midnight, respectively.

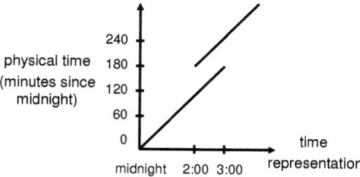

Figure 12.3 Glitches in time representation when switching from daylight saving time to wintertime

Any calculations on time representation that include this specific hour would be incorrect without any additional measures. Linearity in time representation is therefore a necessary feature. A time representation that almost fully meets this requirement is *UTC*, which stands for *Coordinated Universal Time*. One day in UTC has 86400 seconds (with a few exceptions that are explained below), it is a realization of Universal Time (UT), a standard time that divides the world into time zones; time zones maintain a time that has a constant offset to Universal Time. UTC is a combination of atomic time, which is used to set the rate of UTC clocks (i.e. the duration of one second) and the epoch of Universal Time. This synchronization to the UT epoch is the cause of a small deviance of the previously required monotonous representation of time: since UT has been introduced, the rotation of Earth has slowly decelerated, meaning that the average duration of a solar day has slightly increased. Therefore UTC has to be adjusted by introducing a *leap second* every once in a while to keep synchronized with Earth's rotation. A leap second is always inserted as the last second of a day. The time around midnight of such a day is represented in UTC as 23:59:58, 23:59:59, 23:59:60, 00:00:00. While this representation does not yet violate monotony, the UNIX representation, which is shown in the following, will do so.

The UNIX operating system has defined the *UNIX epoch* (also called *POSIX time*). It defines points in time (instants) and uses a reference point, which is January 1st, 1970, 00:00:00 GMT, the start of the UNIX epoch. UNIX time is based on UTC, thus ensuring monotony of time representation – with the exception of the leap second problem. Time is mapped to a number in UNIX time, which starts with zero at the beginning of the epoch and is increased by one every second. According to UTC, most days have 86400 seconds, with a few having 86401 seconds[62]. Still every day increases the number by exactly 86400. January 1st, 2000, 10957 days after January 1st, 1970 is represented by the number 10957*86400 = 946684800.

Ambiguity starts where leap seconds are necessary. While the UNIX time number on one side is used to count 86401 seconds of the day with the leap second, it will repeat in the second after midnight. Therefore, UNIX time is ambiguous for one second

---

[62] Days with one second less than the average day – 86399 seconds – are also possible, but have until now never been used.

whenever a leap second is introduced, which happens about every second year, depending on the changes in Earth rotation. This tradeoff appears acceptable with the benefit of having time that remains synchronized with the changes of day and night we are used to.

The UNIX operating system has defined a type `time_t` for the C programming language. This type commonly uses a signed 32 bit integer number to represent UTC time. Thus, it allows negative numbers for instants before January 1st, 1970 and it has a resolution of one second. An interesting side effect of the 32 bit type is that it will reach its highest possible representation on January 19, 2038, 03:14:08 UTC and will overflow. Problems similar to the ones on New Year of 2000 are expected, where two digit representation of the year (which have been common in early days of computing) have overflown. A solution would be to substitute `time_t` to a 64 bit integer.

Using this type for time representation lacks the required sub-second precision. Aside of that it meets all of the above introduced requirements. To extend precision towards sub-second resolution, different solutions exist. There are specifications for two additional types in C `timeval` and `timespec`. Both of them are structs containing a variable for common UTC time representation and an additional variable for microseconds (in `timeval`) or nanoseconds (in `timespec`), respectively. These extensions allow to continue using the common UTC time representation and operate on both variables, when higher precision is required. However, basic operations like subtraction become more complicated, since it is necessary to respect the carry between the variable representing the fraction of a second and the second variable.

Therefore a different storage method for UTC time is used, which can be found in the time in Java: time in Java is based on UTC and similar to `time_t` in C, only that it uses a long variable (which is defined as a signed 64 bit integer in Java) and defines the number to be interpreted as milliseconds since January 1st, 1970, 00:00:00 GMT. This definition provides sufficient sub-second precision, is almost completely monotonous, it is convenient to operate on, since it is contained in one variable. It is portable, since Java is portable and other modern programming languages are able to process 64 bit numbers (it can also be stored in a database). By using the definition that is used in Java, it also follows a widely adopted definition.

One addition is required, which is a representation for an undefined point in future; an analogy in number representations would be infinity. This representation is needed to indicate that something has started, but not finished yet, for example, a symbol that has been created and still exists. The timestamp for the end needs to contain a number and this number shall be interpreted as "infinity". Therefore this representation is defined to be the highest number that can be represented with a 64 bit signed integer number, which is $(2^{63} - 1)$, or approximately $9.22*10^{18}$ (0x7FFFFFFFFFFFFFFF in hexadecimal representation).

12.5.3 Time Synchronization

The system design foresees the modules to run on different computer, which are connected over a network. The simulator and visualisator, for example, can be started in multiple instances, where some may be on the same computer and others run on computers connected by network. Modules of the system need to interact with other modules or the simulator as well, for example the modules that create and update symbols (these modules need to inform the visualization part of the simulator where to display symbols). Therefore, it is necessary to use proper time synchronization between the different modules and between different instances of the simulator.

Today's computers are equipped with reasonably precise real-time clocks that are able to provide time with the required accuracy over an extended period of time. Still it is necessary to ensure that the deviation between time bases on different computers does not exceed the limits. Therefore, means for synchronizing the real-time clocks on a regular basis are necessary. The common way to do so is to use a timeserver and a protocol that allows adjusting time. In [RFC2030] the Simple Network Time Protocol (SNTP) is defined, a protocol that is commonly used to synchronize clocks using an IP-based network. The accuracy that can be achieved using this protocol (and its predecessor, NTP version 3, which is still in use) is about 1 to 50 milliseconds, depending on the underlying network. This accuracy is sufficient for the tasks of the system. Therefore, the only measure to be taken is to have a client running on each computer that synchronizes regularly with one of the available timeservers.

Another issue is simulation, which uses a different time base. If time is to be slowed down or simply modified to study the behavior of the system in more detail, the modules still need to stay synchronized. Therefore, there needs to be a central instance as a time source that informs all modules about the current time and when to proceed in time. In case of simulation with a virtual time base this time source substitutes the real-time clock and informs all modules when to proceed to the next time slot. Measures have to be taken in the framework for the modules that time can conveniently be switched between real-time and virtual time without the application having to take care about it.

13 Visualization, Simulation and Real-World Installation

An introduction into the concept of the simulation and visualization components is given in [Roe04] and [Har05]. Figure 13.1 shows an overview of the communication paths between the components. The components are described in the following sections. The descriptions focus on the visualization of a virtual environment and the sensor and symbol information that is integrated into visualization.

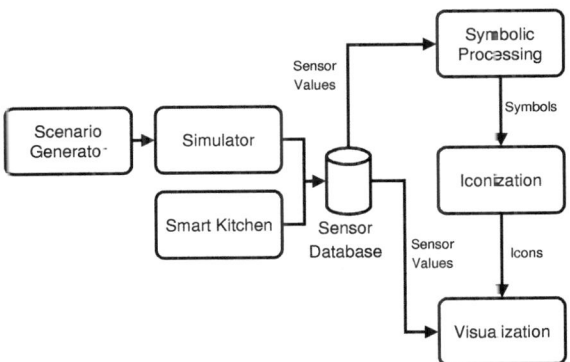

Figure 13.1 Communication paths for simulation and visualization

13.1 Visualization

The visualization component (called *envSim*) displays a three-dimensional (and optionally two-dimensional) simulated environment. It renders and animates three-dimensional objects and creates different views of the environment. Visualization is based on the three-dimensional model of the environment (not shown in Figure 13.1), which is used to display all parts of the environment that are not subject to sensory perception. The basic rendering of the environment can be extended in two ways: by showing sensor information or by showing symbolization information; both ways are configurable in the simulator. When showing sensor information, the basic view of the environment is overlaid with a visualization of important sensor reactions to changes in the environment. For example, a person passing a light barrier causes the light barrier to be triggered. This is visualized, so that a human user can see the reactions of sensors. Some sensors like the light barrier are only triggered briefly, which means that they are *events*; other sensors display their status, which is done permanently (e.g. a temperature sensor, which displays the temperature it measures).

Similar to the sensor view there is also a symbolization view, where the environment is overlaid with information that is created by the symbolization modules. For example, a person that enters a room causes several sensors to be triggered (door contacts, light barriers, tactile sensors in the floor, motion detectors). Again, there are event-like symbols that pop up only briefly and there are symbols that have a permanent status, for example, the symbol for a person: it is permanently displayed and its status (e.g. its position) is updated.

Symbols need to have a graphical representation, which is done by the *Iconization* component (Figure 13.1). Using the SymbolNet framework this component receives all symbols (and updates) that are created by the symbolic modules (shown as one reduced module in Figure 13.1, see also section 10.1[63]). Symbols and their properties are converted into graphical representations; Figure 13.2 shows a selection of icons that are used to display symbols.

Figure 13.2 Selected icons for symbol visualization

Both sensor and symbolic information can be filtered, meaning that the user has the possibility to select the information that shall be displayed. Sensor value display can be limited so that only a subclass of sensors is displayed (e.g. restricted to an area or restricted to certain types of sensors). Symbols and the according icons can also be limited, so that, for example, only symbols on representation level, but not snapshot symbols or microsymbols, are shown.

Visualization is not only used for the virtual environment described in section 13.2, but also to show how the symbolization operates on real-world data. This data originates from the Smart Kitchen installation (section 13.3). A three-dimensional model of the real room is used for visualization (thus the real room is "virtualized"), this way, the same visualization can be used to show sensor values and symbols for the real-world installation and the simulated environment.

13.2 Simulation

The virtual environment that is visualized is intended to be used as an enhancement of a real-world installation; it consists of a three-dimensional layout of rooms and halls, in which virtual sensors are mounted and virtual persons can move. The behavior of the simulation is controlled by the scenario generator (section 13.5).

The data that result from the simulation are fed into the sensor database. This way it can be reused later, either for tuning module behavior or for visualization in virtual time (see section 13.3): depending on the physical model and its complexity there may be considerable effort necessary to derive changes and interactions in the environment. The symbolization module also may take considerable time to achieve a result. Therefore, a simulation run may take more time than the real time that passes during the simulation. To circumvent delays in display, the two parts can be separated: first, simulation is finished, and then it is visualized.

---

[63] Three-dimensional visualization is only done for symbols of the outer world; the maintenance status of the inner world is done in a reduced, textual form.

13.3 Smart Kitchen – A Real-World Installation

The Smart Kitchen is a room at the Institute of Computer Technology that has already been used for earlier works to create a sensor-enabled environment [Rus03]. Since then it has been extended by different types of sensors. Figure 13.3 shows the installation of floor sensors in the Smart Kitchen.

Figure 13.3 Floor sensors in Smart Kitchen

The floor sensors deliver information about the position of persons in the room. It is expected that future technologies like the Thinking Carpet [Vor05] will continue this concept to increase the number of sensors in buildings. Together with motion detectors and light barriers, it is possible to obtain diverse and redundant information to obtain a robust information source of environment variables.

13.4 Virtual Time and Real Time

The visualization component is needed for two different purposes: it shall display a virtual environment with sensors and persons that need to be simulated, and it shall display sensor activity and symbolic processing of the Smart Kitchen installation. While it is important for the real-world installation to react in real-time and display current changes of sensor values together with the according changes in symbolization, the virtual environment has different requirements to timing. When simulation of sensor reactions becomes too complex, the simulation run can take considerable time and therefore cannot be visualized in real-time. The solution is to write the results of simulation into the sensor database and let visualization display the data afterwards. This yields another possibility for displaying the speed of time can be changed for visualization. Important events can be shown at, for example, half the speed; or a whole day can be compressed into a few minutes. Virtual time is a common feature in simulation; in this system it can be used for the virtual environment as well as for the Smart Kitchen: once the sensor data have been created, they can then be replayed from the sensor database.

13.5 Simulation Run Generator

A simulation run is triggered by the simulation run generator (Figure 13.1): first, there is an initialization phase, in which the layout of the virtual environment together with the position for all objects is defined. All sensors and actuators are defined and initialized in this phase as well. The definition of the initialization is called a *setup*. When initialization is finished, the simulation run generator sends updates for different objects in the environment[64]. The updates may be new positions, changes in properties or creation of new objects. This mechanism is used for creating a complete course of events; it is network transparent, meaning that the simulation run generator can run on a different machine than the simulator.

Using the simulation run generator it is possible to create a formal description that defines the movements, updates and creation of all objects (e.g. the behavior of people in an office during a whole day). This description is loaded into the simulation run generator, which then feeds it step by step into the simulator.

13.6 Physical Model of Simulator

The physical model that drives the simulator contains a basic implementation that covers necessary physical effect, like light barriers that are triggered, when a person passes by. The model also considers the physical dimensions of objects and detect collisions between object.

Cameras are sensors that need to be handled specially in the simulator: it is not reasonable to render a complete camera image that is afterwards analyzed by image recognition, because it would not yield proper results. Therefore, the preprocessing is already done in the virtual camera. Instead of providing sensor values, the virtual cameras create symbols. This behavior bypasses symbolic processing and is fed into the iconization module using SymbolNet.

Aside of sensors the simulator also has to animate the actuators like sliding doors, heating, light, and so on. Collisions between actuators and objects need to be handled, which might have impact on the simulation run generator and on the symbolization module.

13.7 Strategic Planning

Another important application of the simulator is to use it for symbolization as a separate module. When the system has perceived a situation, it can consider starting an action. However, before this action is executed, it has to know about the outcome of it. One way that this can be done is to "try it out mentally": the system uses an isolated instance of the simulator to create the start situation; then it starts the action, looks at the outcome and judges, if it shall actually execute the action. Only on

---

[64] In this context, simulated persons are also referred to as objects, because the simulation has to animate both persons and object.

positive results will this action actually be fed into the simulator. The best way to try out actions without impact is to use a separate "mental simulator". This would be a possibility for future extension of the symbolization modules and the simulator.

14 Conclusion

This thesis describes a symbolic data processing model based on both real-world and simulated (virtual world) data. The grounded symbolic model that has been designed demonstrates to be an applicable adaptation of the neuroscientific models that were introduced. This conclusion reviews the output of this work and leads to the outlook, which briefly describes how research can continue based on the work that has been done so far.

An important issue of symbolic processing of information is addressed in this work: while it is state-of-the-art technology to use symbols and apply operations according to a set of rules on these symbols, the symbol system commonly suffers from a lack of meaning, in the sense that the system does not have means for connecting symbols to real-world concepts because the meaning of symbols is not contained in the system. While it can operate on symbols and produce results that may appear reasonable, it does not have intrinsic understanding of the concepts it operates on. By connecting the system's symbols to sensor data, they are grounded to the real world, thus avoiding a purely symbolic system that lacks the connection to a meaning of the symbols. Sensor data is a reflection of real-world influences on the system; by building the symbolic alphabet on sensor data, the bridge between symbolic representation and the outside world is built. A possibility to connect real-world information with symbolic concepts is shown in this work by taking the step from sensor data to microsymbols. By observing the impact of actions to the environment and thus the representation of the outer world, the system is connected to the real world, although it operates only on symbols.

Being mainly a passive, observing system, the sensor data travel from sensor level over multi-layer symbolic processing modules to create a representation of the outer world. Additionally to the representation of the outer world, a representation of the inner world is created to represent system status and maintenance information. The symbol alphabet is engineered by directly applying human knowledge from the domain of building automation, allowing the system to operate on human-understandable concepts and thus creating a transparent model that can be tweaked and tuned by a human operator.

The selected reference applications have proven to provide a broad basis of different possible scenarios. Thus, it is prepared for future extension and refinements of scenario recognition. Due to the modular design of the system which operates transparently over the network, it is possible to observe the system closely while it operates. This is an intended analogy to neurobiological sciences, which partly rely on the study of brain activities. Naturally, a human-crafted technical system is easier to observe; still, the network transparency allows visualizing data flow between modules and forces modular design with strictly defined interfaces. As an additional effect, the network transparent modularity of the system allows for convenient extension of system functionality. This means that a single component is bound to work on a single computer, while the sum of all components can easily be distributed over a set of networked computers, allowing to connect more resources upon introduction of new necessary system components.

The condensing of information in a multi-layered process produces a scalable system design that follows the concepts that were extracted from the neuropsychological domain. By using a lot of sensor information originating from redundant and diverse sources, it is conceptually possible to maintain a consistent view of the world. While the lower layers need to process incoming information efficiently in parallel, higher-level processing can rely on condensed data, which has been preprocessed and thus represents information that is more expressive.

Symbolic representation of objects and events of the real world are not a novel concept. However, the symbolization of physical variables like time and motion, which is taken from human understanding of such variables, is an approach that does not follow the engineering method of measuring a quantity in numerical values and operating on these. On the highest level of symbolization (the representation level), this is comparable to an emotion that is, for example, linked to an object and its speed.

The three-layer symbolic processing model that is introduced in this work has emerged from different requirements. First, a system design with strictly hierarchical communication structure allows for clear modularization and avoids error prone cross-linked communication paths. On the other hand, information exchange in the human brain is complex and – on higher cognitive levels – not hierarchical at all. Thus, the symbolization model designed here represents the common denominator of both models: symbols on lower levels are combined with others to create symbols on higher levels. Each level is a representation of information, and the combination of symbols creates a new representation of information. Following this approach, the system creates representations of representations, especially on higher symbolic levels, where symbols can be combined to create symbols on the same layer. Researchers use representations of representations as one model to describe the operation of the cortex: representations from different functional systems are used to create new representations, which again are used for even other representations.

The technology used in this work has the potential to increase comfort and safety of its users, but it may also be abused for malicious surveillance activities, which would compromise the individual's privacy. All the more this work shall be seen as a contribution to prevent malicious use of surveillance that results from the lack of understanding of underlying technology. If, for example, cameras are used to observe persons and the cameras provide visual information, the risk to compromise privacy is much higher than if symbolizing visual sensor information, which extracts relevant information. By communicating only relevant information in symbolized format, much of the compromising information (which is in fact irrelevant for the application of the system) is stripped off and thus not available. Symbolization is therefore a means for protection of privacy. Nevertheless, the operator of such a system has to take measures to prevent the leaking of personal information, which can only be solved by an appropriate infrastructure and not by technology itself.

15 Outlook

The design for a system that behaves similar to the abilities of a human mind raises an interesting thought: the human mind is intrinsically bound to its body and in particular the brain tissues that builds the center of all higher cognitive functions. It is today well known that body and mind are tightly interwoven and one influences the development and abilities of the other. However, when the mind can be engineered as "a piece of software", the link between body and mind is not given any more. Software is also bound to hardware (just like the mind is bound to the body), but if we look at the development of computer hardware over time, we see that hardware can change without strongly affecting software. Programs were executed on computers made up of relays, which were then replaced by transistors, followed by integrated circuits. The properties of the silicon that formed the transistors did not affect the software. Furthermore, the most important ability of the human body – its ability to interact with its environment – is only given in a very limited way when looking at computers. As long as there is only a limited interconnection between a machine and the environment it operates in, we can think of the mind in the machine as a disembodied mind, meaning that whatever happens in the program is largely not dependent on the physical hardware it uses as the platform for execution. Future developments in this field will show how a mind can evolve without the link to a physical body or, as another possible outcome, it will show that intelligent behavior always requires interaction with the outer world to be able to develop in the first place.

An ability that has deliberately been left out in this work is the ability to learn and adapt to changes in the outer world. The complexity that arises when the system changes its behavior dynamically, depending on how much data it was able to process and learn, is too much of a challenge for a new model like the AFS model. The next steps therefore need to focus on making single modules more flexible by introducing learning abilities. The model can remain unchanged, but separated parts can be enhanced and improved.

Lurija has stated that human perception requires complex coding of incoming information and that this coding is closely linked to language [Lur01]. He further states that perception cannot happen without language. What has been done in this work is to create a crude language that enables perception in the way Lurija (and others) intended it: far away from the complexity of human spoken or written language, but close enough to the "inner language" of a building automation system that it can be used to process information to create a world representation. What needs to be done now is to increase the complexity of the symbol alphabet, to create a lot more symbols and let the system operate with it. However, this will in the long run not be possible by using "hand-crafted" symbols, but will require methods for automatic symbol generation. Such methods exist and yield hope to finally create a system that is able to use symbols that are grounded to real-world information on the one hand and have a human-understandable meaning on the other hand.

Bibliography

[Ach00]   A. Achilles. SQL: standardisierte Datenbanksprache vom PC bis zum Mainframe. 7th edition, Oldenbourg, Munich/Vienna, 2000

[And95]   J. Anderson. An introduction to neural networks. MIT Press, Cambridge, Massachusetts, 1995

[ASN02]   International Organization for Standardization. Abstract Syntax Notation One (ASN.1): Specification of basic notation (ISO/IEC 8824-1:2002), Geneva, Switzerland, 2002

[Bab14]   J. Babinski. Contribution a l'étude des troubles mentaux dan l'hémiplégie organique cérébrale (anosognosie). Revue neurologique 27, 1914

[Bec94]   A. Bechara, A. R. Damasio, H. Damasio, S. W. Anderson. Insensitivity to future consequences following damage to human prefrontal cortex. Cognition, 50. p. 7 – 15

[Bra04]   E. Brainin, D. Dietrich, P. Palensky, C. Rösener. Neuro-bionic Architecture of Automation Systems - Obstacles and Challenges. In Proceedings of the 7th IEEE Africon Conference, Gaborone, Botswana, 2004

[Bro86]   R. A. Brooks. A robust layered control system for a mobile robot. IEEE Journal of Robotics and Automation, RA-2, April 1986

[Dam94]   A. Damasio. Descartes' Error. Emotion, reason, and the human brain. Penguin Books Ltd., London, England, 1994

[Dam99]   A. Damasio. The Feeling of What Happens. Body and Emotion in the Making of Consciousness. Harcourt Brace & Company, New York, 1999

[DER02]   International Organization for Standardization. ASN.1 encoding rules: Specification of Basic Encoding Rules (BER), Canonical Encoding Rules (CER) and Distinguished Encoding Rules (DER) (ISO/IEC 8825-1:2002), Geneva, Switzerland, 2002

[Die02]   D. Dietrich, C. Tamarit, G. Russ. Bionische Modellierung. Elektronik Report 12, Vienna, 2002

[Die04]   D. Dietrich, W. Kastner, Th. Maly, Ch. Rösener, G. Russ, H. Schweinzer. Situation Modeling. In Proceedings of the 5th IEEE International Workshop on Factory Communication Systems (WFCS 2004), Vienna, Austria, 2004

[Die04b]   D. Dietrich, W. Kastner, H. Schweinzer: Wahrnehmungsbewusstsein in der Automation - ein "bionischer" Denkansatz; at – Automatisierungstechnik 52 (2004) 3, Oldenbourg Verlag

[Ech94]   Echelon Corporation: LonTalk Protocol Specification. Version 3.0, Document No. 19550, United States of America, 1994.

[Ech95]   Echelon Corporation: Neuron C Programmer's Guide. Revision 4, Document No. 29300, United States of America, 1995.

[Elm02]   W. Elmenreich. Sensor Fusion in Time-Triggered Systems. Dissertation an der Fakultät für Elektrotechnik der Technischen Universität Wien, 2002

[Enc04]   Encyclopædia Britannica, from Encyclopædia Britannica Ultimate Reference Suite 2004 DVD. Copyright © 1994-2003 Encyclopædia Britannica, Inc. May 30, 2003

[Fod88]   J. Fodor. Modules, frames, fridgeons, sleeping dogs and the music of the spheres. In: The robot's dilemma: the frame problem in artificial intelligence. 2. print, ed. by Z. Pylyshyn. Ablex Publ. Co., Norwood, NJ, 1988

[Fre74]   S. Freud. Civilization and Its Discontent, S.E., 21:59, 1930, pp. 71-72 in The Standard Edition of the Complete Psychological Works of Sigmund Freud (SE) 24 Volumes, ed. by James Strachey et al. The Hogart Press and the Institute of Psycho-Analysis, London, 1974

[Gol02]   E. Goldstein. Wahrnehmungspsychologie - Eine Einführung. 2. Auflage, Spektrum Akademischer Verlag, November 2002

[Hac04]   Hacke, A., Sowa, M., Der Weiße Neger Wumbaba. Kleines Handbuch des Verhörens. Verlag Antje Kunstmann GmbH, München, 2004

[Har05]   H. Hareter, G. Pratl, D. Bruckner. Simulation and Visualization System for Sensor and Actuator Data Generation. In Proceedings of 6th IFAC International Conference on Fieldbus Systems and their Applications (FET 2005), Puebla, Mexico, 2005, p. 56 – 63

[Har90]   S. Harnad. The Symbol Grounding Problem. Proceedings of the ninth annual international conference of the Center for Nonlinear Studies on Self-organizing, Collective, and Cooperative Phenomena in Natural and Artificial Computing Networks on Emergent computation, Los Alamos, New Mexico, United States, 1990, pp. 335 – 346

[Hau95]   K. Hauser. A. Bittleston, A.D. Gibson, W.B. Forward *Kaspar Hauser Speaks for Himself: Kaspar's Own Writings.* Camp Hill Press, 1995

[IEEE96]   IEEE Standard for a High Performance Serial Bus (IEEE1394-1995). ISBN 1-55937-583-3, IEEE, 1996.

[IEEE00]   IEEE Computer Society: IEEE Standard for a High Performance Serial Bus - Amendment 1, 2000

[ISO8601] ISO 8601:2004 Data elements and interchange formats - Information interchange - Representation of dates and times. Third edition, International Organization for Standardization, 2004

[Kap00]   K. Kaplan-Solms, M. Solms. Clinical Studies in Neuro-Psychoanalysis. International Universities Press, Inc., Madison, CT, 2000

[Lak05]   B. Lakotta. Die Natur der Seele. Hatte Sigmund Freud doch recht? Der Spiegel 16/2005, dated Apr. 18th, 2005

[Lob02]   M. Lobashov, G. Pratl, T. Sauter. Applicability of Internet Protocols for Fieldbus Access. Proceedings of the IEEE International Workshop on Factory Communication Systems (WFCS 2002), Västeraas, Sweden, August 2002

[Lob03]   M. Lobashov, G. Pratl, T. Sauter. Implications of Power-line Communication on Distributed Data Acquisition and Control System. In Proceedings of Emerging Technologies and Factory Automation (ETFA'03), Lisbon, Portugal, Sept. 2003

[Lob05]   M. Lobashov. Applicability of Internet Protocols to Remote Fieldbus Access. Dissertation an der Fakultät für Elektrotechnik der Technischen Universität Wien, 2005.

[Lon02]   D. Dietrich, K. Kabitzsch, G. Pratl (editors). LonWorks – Gewerkeübergreifende Systeme. VDE-Verlag, Germany, 2002

[Lur01]   A. Lurija. Das Gehirn in Aktion. Rowohlt Taschenbuch Verlag, Reinbek bei Hamburg, 2001

[Mah04] S. Mahlknecht. Energy-Self-Sufficient Wireless Sensor Networks for Home and Building Environment. Ph.D. Thesis, Vienna University of Technology, 2004

[McC69] J. McCarthy, P. J. Hayes. Some Philosophical Problems from the Standpoint of Artificial Intelligence. Machine Intelligence 4, Edinburgh University Press, pp. 463-502, 1969

[Nus04] C. Nüsslein-Volhard. Das Werden des Lebens. C.H. Beck oHG, 2004

[Pal03] P. Palensky, P. Rössler, D. Dietrich. Heim- und Gebäudeautomatisierung zur Effizienzsteigerung in Gebäuden. Elektrotechnik und Informationstechnik (e&i), 4, 2003

[Pen50] W. Penfield, T. Rasmussen. The Cerebral Cortex of Man. A Clinical Study of Localization of Function. New York, The Macmillan Comp., 1950

[Pez05] L. Pezawas, A. Meyer-Lindenberg, E. M. Drabant, B. A. Verchinski, K. E. Munoz, B. S. Kolachana, M. F. Egan, V. S. Mattay, A. R. Hariri, D. R. Weinberger. 5-HTTLPR polymorphism impacts human cingulate-amygdala interactions: a genetic susceptibility mechanism for depression, Nature Neuroscience 8, May 2005, p. 828 – 834

[Pfe87] T. Pfeifer, K.-U. Heiler. Ziele und Anwendungen von Feldbussystemen, Automatisierungstechnische Praxis, Vol. 29, 1987, pp. 549 – 557

[Pra01] G. Pratl, M. Lobashov, T. Sauter. Highly modular Gateway Architecture for Fieldbus/Internet Connections", Proceedings of Fieldbus Systems and their Applications (FeT 2001), Nancy, France, November 2001

[Pra05a] G. Pratl, P. Palensky. Project ARS – The next step towards an intelligent environment. Proceedings of the IEE International Workshop on Intelligent Environments, 2005, p. 55 – 62.

[Pra05b] G. Pratl, W. T. Penzhorn, D. Dietrich, and W. Burgstaller. Perceptive awareness in building automation. In 3rd International Conference on Computational Cybernetics (ICCC'05) Conference Proceedings, Mauritius 2005

[Rad05] R. Radke, S. Andra, O. Al-Kofahi, and B. Roysam, "Image change detection algorithms: A systematic survey", IEEE Transactions on Image Processing, Vol. 14, pp. 294–307, 2005.

[Roe04]   Ch. Rösener, H. Hareter, W. Burgstaller, G. Pratl. Environment Simulation for Scenario Perception Models. In Proceedings of the 5th IEEE Workshop on Factory Communication Systems (WFCS 2004), Vienna, Austria, 2004

[Rud03]   M. Rudberg, J. Olhager, "Manufacturing networks and supply chains: an operations strategy perspective", Omega, vol. 31, 2003

[Rus03]   G.Russ. Situation-dependent behavior in building automation. Dissertation an der Fakultät für Elektrotechnik der Technischen Universität Wien, 2003

[Russel03] S. Russell, P. Norvig. Artificial Intelligence: A Modern Approach. Englewood Cliffs, New Jersey: Prentice Hall, 2nd edition, 2003

[Ryd04]   D. Ryder SINBAD Neurosemantics: A Theory of Mental Representation. Mind & Language, Vol 19. Issue 2, p 211, April 2004

[Sac87]   O. Sacks. The Man Who Mistook His Wife For a Hat. Summit Books/Simon & Schuster, Inc., New York, 1987

[Sal05]   B. Sallans, D. Bruckner, G. Russ. Statistical Model-Based Sensor Diagnostics for Automation Systems. In Proceedings of 6th IFAC International Conference on Fieldbus Systems and their Applications (FET 2005). Puebla, Mexico, 2005, p. 239 – 246.

[Saut02] T. Sauter, M. Lobashov, G. Pratl. Lessons Learnt From Internet Access to Fieldbus Gateways. Proceedings of IECON'02, Sevilla, Spain, November 2002

[Sea80]   J. Searle. Minds, Brains, and Programs. Behavioral and Brain Sciences 3, p. 417 – 424, 1980

[Sol02]   M. Solms and O. Turnbull. The brain and the inner world. Other Press LLC, New York, 2002.

[Tam03]   C. Tamarit Fuertes. Automation system perception. Dissertation an der Fakultät für Elektrotechnik der Technischen Universität Wien, 2003

[Wan92]   Y. Wang, B. Frost. Time to collision is signalled by neurons in the nucleus rotundus of pigeons. Nature, 1992, 356(6366), p. 236-238

[Wol05]   F. Wolf. Symmetry, Multistability, and Long-Range Interactions in Brain Development. Physical Review Letters 95 Nr. 20, 208701,

American Physical Society, Nov. 7th, 2005

[Yu04]   C. Yu, D. Ballard. A multimodal learning interface for grounding spoken language in sensory perceptions. ACM Transactions on Applied Perception, Volume 1, Issue 1, July 2004, p. 57 – 80

[Zad65] L. A. Zadeh, Fuzzy Sets, Information and Control, 1965

Internet Links

[Blue04]   Blue Brain Project. http://bluebrainproject.epfl.ch, September 14th, 2004, accessed on March 14th, 2006

[Max02]   Max Planck Institute for Biological Cybernetics. Visual Illusions: What we see is what we expect to see. May 13th, 2002, http://www.kyb.mpg.de/bu/demo/index.html, accessed on Jan 21st, 2006

[Neu06]   O. Turnbull, Y. Yovell. Neuro-Psychoanalysis. http://www.neuro-psa.org, accessed on Jan 25th, 2006

[Open06]   OpenCV Library Wiki. http://opencvlibrary.sourceforge.net, accessed on March 30th, 2006.

[Pet97]   I. Peterson. Ivars Peterson's MathLand http://www.maa.org/mathland/mathland_5_12.html, May 12th 1997, accessed on March 14th, 2006.

[RFC793]   J. Postel. Transmission Control Protocol, RFC 793. Sept. 1981, http://www.rfc-editor.org, accessed on Jan 19th, 2006

[RFC1157] J. Case, M. Fedor, M. Schoffstall, J. Davin. A Simple Network Management Protocol (SNMP), RFC 1157. May 1990, http://www.rfc-editor.org, accessed on Jan 19th, 2006

[RFC1945] T. Berners-Lee, R. Fielding, and H. Frystyk. Hypertext Transfer Protocol – HTTP/1.0, RFC 1945. May 1996, http://www.rfc-editor.org, accessed on Jan 19th, 2006

[RFC2030] D. Mills. Simple Network Time Protocol (SNTP) Version 4 for IPv4, IPv6 and OSI, RFC 2030. Oct. 1996, http://www.rfc-editor.org, accessed on Jan 19th, 2006

[RFC2251] M. Wahl, T. Howes, S. Kille. Lightweight Directory Access Protocol (v3), RFC 2251. Dec 1997, http://www.rfc-editor.org, accessed on Jan 19th, 2006

[Sha04]   M. Shanahan. The Frame Problem. Feb. 23rd, 2004, http://plato.stanford.edu/entries/frame-problem, accessed on Jan 19th, 2006

[Vor05]   Vorwerk & Co. Teppichwerke. Zukunft braucht Visionen. http://www.vorwerk-teppich.de/sc/vorwerk/template/Thinking_Carpet_deu.html, accessed on Sep 2nd, 2005

Die VDM Verlagsservicegesellschaft sucht für wissenschaftliche Verlage abgeschlossene und herausragende

Dissertationen, Habilitationen, Diplomarbeiten, Master Theses, Magisterarbeiten usw.

für die kostenlose Publikation als Fachbuch.

Sie verfügen über eine Arbeit, die hohen inhaltlichen und formalen Ansprüchen genügt, und haben Interesse an einer honorarvergüteten Publikation?

Dann senden Sie bitte erste Informationen über sich und Ihre Arbeit per Email an *info@vdm-vsg.de*.

Sie erhalten kurzfristig unser Feedback!

VDM Verlagsservicegesellschaft mbH
Dudweiler Landstr. 99
D - 66123 Saarbrücken

Telefon +49 681 3720 174
Fax   +49 681 3720 1749

www.vdm-vsg.de

Die VDM Verlagsservicegesellschaft mbH vertritt

Printed by Books on Demand GmbH, Norderstedt / Germany